The
LAST DAYS
of the
DINOSAURS

ALSO BY RILEY BLACK

FOR ADULTS
Written in Stone
My Beloved Brontosaurus
Skeleton Keys
Deep Time

FOR CHILDREN
Prehistoric Predators
Did You See That Dinosaur?

The
LAST DAYS
of the
DINOSAURS

An Asteroid, Extinction,
and the Beginning of Our World

RILEY BLACK

ST. MARTIN'S
PRESS
NEW YORK

First published in the United States by St. Martin's Press,
an imprint of St. Martin's Publishing Group

THE LAST DAYS OF THE DINOSAURS. Copyright © 2022 by Riley Black.
All rights reserved. Printed in the United States of America.
For information, address St. Martin's Publishing Group,
120 Broadway, New York, NY 10271.

www.stmartins.com

Designed by Steven Seighman
Illustrations by Kory Bing

Library of Congress Cataloging-in-Publication Data

Names: Black, Riley, author.
Title: The last days of the dinosaurs : an asteroid, extinction, and the
    beginning of our world / Riley Black.
Description: First edition. | New York : St. Martin's Press, 2022. |
    Includes bibliographical references.
Identifiers: LCCN 2021052246 | ISBN 9781250271044 (hardcover) |
    ISBN 9781250271051 (ebook)
Subjects: LCSH: Extinction (Biology) | Dinosaurs—Extinction.
Classification: LCC QE721.2.E97 B57 2022 | DDC 576.8/4—dc23/
    eng/20211208
LC record available at https://lccn.loc.gov/2021052246

Our books may be purchased in bulk for promotional, educational, or busi-
ness use. Please contact your local bookseller or the Macmillan Corporate
and Premium Sales Department at 1-800-221-7945, extension 5442, or by
email at MacmillanSpecialMarkets@macmillan.com.

First Edition: 2022

10 9 8 7 6 5 4 3 2 1

For Margarita
No amount of time would have been enough.

Contents

Preface

Catastrophe is never convenient.

The dinosaurs never expected it. Nor did any of the other organisms, from the tiniest bacteria to the great flying reptiles of the air that were thriving on a perfectly normal Cretaceous day 66 million years ago. One moment life, death, and renewal proceeded just as they had the day before, and the day before that, and the day before that, stretching back through millions upon millions of years. The next, our planet suffered the worst single day in the entire history of life on Earth.

In an instant, life's entangled bank was thrown into fiery disarray. There were no warning signs, no primordial klaxon that would blare and send Earth's organisms rushing to whatever refuges they might find. There was no way for any species to prepare for the disaster that came crashing down from the sky with an explosive force 10 billion times greater than the atomic bombs detonated at the end of World War II. And that was just the beginning. Fires, earthquakes, tsunamis, and the choking hold of an impact-created winter that lasted for years all had their own deadly roles to play in what followed.

The disaster goes by different names. Sometimes it's called the end-Cretaceous mass extinction. For years, it was called the Cretaceous-Tertiary, or K-T, mass extinction that marked the end of the Age of Reptiles and the beginning of the third, Tertiary age of life on Earth. That title was later revised according to the rules of geological arcana to the Cretaceous-Paleogene mass extinction, shortened to K-Pg. But no matter what we call it, the scars in the stone tell the same story. Suddenly, inescapably, life was thrown into a horrible conflagration that reshaped the course of evolution. A chunk of space debris that likely measured more than seven miles across slammed into the planet and kicked off the worst-case scenario for the dinosaurs and all other life on Earth. This was the closest the world has ever come to having its Restart button pressed, a threat so intense that—if not for some fortunate happenstances—it might have returned Earth to a home for single-celled blobs and not much else.

The effects of the impact were swift and dire. The heat, fire, soot, and death blanketed the planet in a matter of hours. What happened at the end of the Cretaceous wasn't a prolonged pulse of die-offs from depleted atmospheric oxygen or acidified seas. This calamity was as immediate and horrific as a bullet wound. The fates of entire species, entire families of organisms, were irrevocably changed in a single moment.

Biologists still argue about what the definition of life truly is—reproduction, growth, movement—but the one amazing fact that we are confronted with every day is that life is incredibly, irrepressibly resilient. Every organism alive today is tied together, each life connected to the one before it. Even as we acknowledge that 99 percent of all species that once lived are now extinct, our world is still brimming with organisms that have survived, evolved, and thrived in their own ways.

In fact, much of our present era owes its existence to the destruction of the K-Pg disaster. The world as we know it today is the continued flowering after a disaster, life not only coming back but reshaped by the very nature of the cataclysm.

In the hours, days, weeks, months, and years following impact, almost every branch in the tree of life was lopped off, damaged, or struggled to grow. Even the organisms that we think of as survivors were not left unscathed. During the K-Pg catastrophe, there were mass extinctions of mammals, lizards, birds, and more, the ecological chaos touching the whole of life on Earth. From the foggy and sometimes dim windows of the fossil record, paleontologists have estimated that about 75 percent of known species that were alive at the end of the Cretaceous were not present in the next sliver of time. As if to drive the point home, a band of clay packed with the metal iridium marks the boundary between the Age of Dinosaurs and the opening chapters of the Age of Mammals. In some places, such as eastern Montana and the western Dakotas, you can follow the story layer by layer, watching the likes of *Triceratops* disappear as a world of diminutive fuzzballs begin to flourish in a new Age of Mammals.

We still feel the loss. As a child, I felt it patently unfair that I could not ride my very own *Tyrannosaurus rex* to school. Even though I've never seen them beyond distorted, permineralized bones, I feel like I miss the non-avian dinosaurs—nostalgia for a time I can never witness, when dinosaurs ruled the Earth. But if the non-avian dinosaurs had survived, our own story would have been altered. Or perhaps prevented altogether.

Not only would mammals have remained small under an extended regime of non-avian dinosaurs, but the earliest, shrewlike primates might have stayed in tight competition with the dominant marsupials. Our ancestors would have been molded in different ways, and it's likely, if not certain, that the world would never have been suitable for a mostly hairless, bipedal ape with a big brain and a penchant for remodeling the planet. The mass extinction at the end of the Cretaceous isn't just the conclusion of the dinosaurs' story, but a critical turning point in our own. We wouldn't exist without the obliterating smack of cosmic rock that plowed itself into the ancient Yucatán. Both stories are present in that moment. The rise and the fall are inextricable.

And here, we often leave the epic tale. The dinosaurs were dominant, even cocky in our prehistoric visions. The largest, strangest, and most ferocious of all inhabited the Late Cretaceous world of soggy swamps and steaming forests. A wayward asteroid suddenly ended their reign, leaving the meek to inherit the Earth. Just as the dinosaurs once benefitted from a mass extinction that allowed them to step out of the shadow of ancient crocodile relatives 201 million years ago, so, too, were our warm-blooded, snuffly little forebears the recipients of good fortune they never earned nor have ever repaid.

We entirely gloss over the nature of recovery, or what made the difference between the survivors and the dead. We obsess over what we lost—blinded to how, even in the shocking cold that followed the initial heat of annihilation, life was already beginning to reseed and recover. It's an extension of how we often cope in the wake of our own personal traumas, remembering the wounds as we struggle to see the growth stimulated by terrible events. Resilience has no meaning without disaster. And that's what led me to this story, the tale of how life suddenly shifted but nevertheless continued to bring us to the here and now. What I'm going to tell you involves hurt and destruction, but that is only the setting for a turning point that's often been taken as a given or somehow inevitable. This is the story of how life bounced back from the worst day in history. Life's losses were sharp and deeply felt 66 million years ago, but each fiddlehead struggling for light, each shivering mammal in its burrow, each turtle that plopped off a log into weed-choked waters set the stage for the world as we know it now. This is not a monument to loss. This is an ode to resilience that can only be seen in the wake of catastrophe.

GEOLOGIC TIMELINE

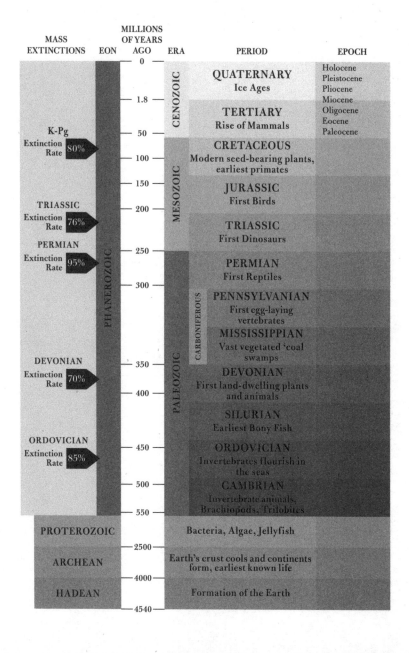

MASS EXTINCTIONS	EON	MILLIONS OF YEARS AGO	ERA	PERIOD	EPOCH
		0	CENOZOIC	**QUATERNARY** Ice Ages	Holocene Pleistocene Pliocene
		1.8			Miocene
				TERTIARY Rise of Mammals	Oligocene Eocene Paleocene
K-Pg Extinction Rate 80%		50			
		100	MESOZOIC	**CRETACEOUS** Modern seed-bearing plants, earliest primates	
		150		**JURASSIC** First Birds	
TRIASSIC Extinction Rate 76%		200		**TRIASSIC** First Dinosaurs	
PERMIAN Extinction Rate 95%	PHANEROZOIC	250		**PERMIAN** First Reptiles	
		300	PALEOZOIC	**PENNSYLVANIAN** First egg-laying vertebrates (CARBONIFEROUS)	
				MISSISSIPPIAN Vast vegetated 'coal swamps	
DEVONIAN Extinction Rate 70%		350		**DEVONIAN** First land-dwelling plants and animals	
		400		**SILURIAN** Earliest Bony Fish	
ORDOVICIAN Extinction Rate 85%		450		**ORDOVICIAN** Invertebrates flourish in the seas	
		500		**CAMBRIAN** Invertebrate animals, Brachiopods, Trilobites	
		550			
PROTEROZOIC				Bacteria, Algae, Jellyfish	
		2500			
ARCHEAN				Earth's crust cools and continents form, earliest known life	
		4000			
HADEAN				Formation of the Earth	
		4540			

Introduction

Picture yourself in the Cretaceous period. It's a day like most any other, a sunny afternoon in the Hell Creek of ancient Montana about 66 million years ago. The ground is a bit mushy, a fetid muck saturated from recent rains that caused a nearby floodplain stream to overrun its banks. If you didn't know any better, you might think you were wading on the edge of a Gulf Coast swamp on a midsummer day. Magnolias and dogwoods shoulder their way into stands of conifers, ferns, and other low-lying plants gently waving in the light breeze drifting over the open ground you now stand upon. But a familiar face soon reminds you that this is a different time.

A *Triceratops horridus* ambles along the edge of the forest, three-foot-long brow horns slightly swaying to and fro as the pudgy dinosaur shuffles its scaly, ten-ton bulk over the damp earth. The dinosaur is a massive quadruped, seemingly a big, tough-skinned platform meant to support a massive head decorated with a shield-like frill jutting from the back of the skull, a long horn over each eye, a short nose horn, and a parrot-like beak great for snipping vegetation that is ground to messy pulp by the plant-eater's cheek teeth. The massive herbivore snorts, making some unseen mammal chitter and scramble in alarm somewhere in the shaded depths of the woods. At this time of the day, with the sun still high and temperatures above 80 degrees, there's barely another dinosaur in sight—the only other "terrible lizards" plainly in view are a couple of birds perched on a gnarled branch peeking out from just inside the shadow of the forest. The avians seem to grin, their tiny insect-snatching teeth jutting from their beaks.

This is where we'll watch the Age of Dinosaurs come crashing to a fiery close.

In a matter of hours, everything before us will be wiped away. Lush verdure will be replaced with fire. Sunny skies will grow dark with soot. Carpets of vegetation will be reduced to ash. Contorted carcasses, dappled with cracked skin, will soon dot the razed landscape. *Tyrannosaurus rex*—the tyrant king—will be toppled from their throne, along with every other species of non-avian dinosaur no matter their size, diet, or disposition. After more than 150 million years of shaping the world's ecosystems and diversifying into an unparalleled saurian menagerie, the terrible lizards will come within a feather's breadth of total annihilation.

We know the birds survive, and even thrive, in the aftermath of what's to come. A small flock of avian species will carry on their family's banner, perched to begin a new chapter of the dinosaurian story that will unfold through tens of millions of years to our modern era. But our favorite dinosaurs in all their toothy, spiked, horned, and clawed glory will vanish in the blink of an eye, leaving behind scraps of skin, feather, and bone that we'll unearth eons later as the only clues to let us know that such fantastic reptiles ever existed. Through such unlikely and delicate preservation our favorite dinosaurs will become creatures that defy tense—their remains still with us, but stripped of their vitality, simultaneously existing in the present and the past.

The non-avian dinosaurs won't be the only creatures to be so harshly cut back. The great, batwinged pterosaurs, some with the same stature as a giraffe, will die. Fliers like *Quetzalcoatlus*, with a wingspan wider than a Cessna and capable of circumnavigating the globe, will disappear just as quickly as the non-avian dinosaurs. In the seas, the quad-paddled, long-necked plesiosaurs and the Komodo dragon cousins called mosasaurs will go extinct, as well as invertebrates like the coil-shelled squid cousins, the ammonites, and flat, reef-building clams bigger than a toilet seat. The diminutive and unprepossessing won't get a pass either. Even among the surviving families of the Cretaceous world, there will be dramatic losses. Marsupial mammals will almost be wiped out in North America, with lizards, snakes, and birds all suffering their own decimation, too. Creatures of the freshwater rivers and ponds will be among the few to get any sort of reprieve. Crocodiles, strange reptilian crocodile mimics called champsosaurs, fish, turtles, and amphibians will be far more resilient in the face of the impending disaster, their lives spared by literal inches.

We know the ecological murder weapon behind this Cretaceous case study. An asteroid or similar body of space rock some seven miles across slammed into Earth, leaving a geologic wound over fifty miles in diameter. Most species from the Cretaceous disappeared in the aftermath. It's difficult to stress the point strongly enough. The loss of the dinosaurs was just the tip of the ecological iceberg. Virtually no environment was left untouched by the extinction, an event so severe that the oceans themselves almost reverted to a soup of single-celled organisms.

We are fearfully enraptured with the idea of such terrible devastation. When the impact at the end of the Cretaceous was scientifically confirmed, news of the disaster inspired not one but two blockbuster films about planet-killing asteroids in the summer of 1998. That such a huge rock could kill more than half of Earth's known species suddenly seemed as obvious as the lethality of a gunshot. Simply knowing the terrible consequences of this disaster has been enough for us to look at the night sky with continued suspicion. If it happened before, it may happen again. NASA keeps an eye on the sky through their Sentry program, hoping to identify threatening asteroids and comets before they get too near.

But we often forget the unusual nature of the K-Pg crisis. Experts have often spoken of the calamity as part of the Big Five—a quintet of mass extinctions that have radically altered life's history. The first extinction crisis, between 455 to 430 million years ago, reshaped the oceans, erasing entire families of archaic invertebrate weirdos and allowing fish to thrive. Rapid global cooling and plummeting sea levels killed about 85 percent of known marine species, reshuffling the evolutionary deck. The second event, spanning 376 to 360 million years ago, shook life up once more. Precisely what caused the disaster is unknown—a drop in ocean oxygen levels is suspected—but the sudden change killed about half of known creatures, reducing the diversity among organisms like trilobites and corals that formed the basis of ancient reefs.

Worse still was the third, peaking about 252 million years

ago. This was the Great Dying, fueled by incomprehensibly violent and sustained volcanic activity that wiped out about 70 percent of known species on both land and sea through climate and atmospheric changes. Our protomammal ancestors, who had held sway in terrestrial ecosystems, were almost entirely extinguished. Their downfall is what allowed reptiles, including dinosaurs, to stage their evolutionary coup. Following that, about 201 million years ago, another disaster killed off a great number of the crocodile relatives that ruled the land and gave dinosaurs their shot at dominance. Once again, intense eruptions were to blame. Greenhouse gases belched into the atmosphere, spurring a burst of global warming followed by intense global cooling. Atmospheric oxygen levels dropped, the seas became more acidic, and the drastic shifts between too hot and too cold were too much for many species to cope with.

But none of these catastrophes were quite like the extinction event that ended the Mesozoic. These previous apocalypses took place over hundreds of thousands or even millions of years, with phenomena like intense volcanic activity and climate change creating grinding, protracted transformations that shifted the makeup of life on Earth over long time spans. The causes of death were also highly variable—ocean acidification prevented shell-building creatures from constructing their calcium carbonate homes, for example, while decreased atmospheric oxygen might have slowly choked terrestrial organisms. What happened at the close of the Cretaceous, however, had global reach. And it happened fast.

The happenstances that triggered the Late Cretaceous extinction culminated in one terrible instant, a rare sliver of time that we can pinpoint as the very moment that life would never be the same. Before the strike, thousands of species flourished on every continent. There were so many varieties of dinosaurs and assorted other creatures that paleontologists are still clocking overtime to find them all, with new toothy, sharp-clawed wonders being named every year. Experts even expect that there were scores of species we'll never know as they lived in places where the circumstances of deposition and sedimentation did not allow them to be preserved, such as dinosaurs that lived in the mountains or other environments that were eroded rather than laid down as layers in stone. Mesozoic life was at its peak. Then, almost overnight, the dinosaurs were all but extinct and the planet's ecosystems were in disarray. This was the worst single day in the history of life on Earth, followed by tens of thousands of years of struggle for the survivors.

Our view of the K-Pg extinction has been hard-won. In fact, the task has involved overcoming our greatest weakness—human hubris. When the famously cantankerous British anatomist Richard Owen coined the name "Dinosauria" in 1842, the great reptiles weren't all that much of a mystery. At the time, only three were known to scientists and the scaly trio seemed to mark part of life's expected progression. Geologists had identified an Age of Fishes, an Age of Reptiles, and an Age of Mammals, moving from low, squishy, squiggling forms of life through scaly monstrosities who were little more than a paleontological sideshow before mammals took up their starring roles. Whether understood as part of a creator's plan or evolution's great march, dinosaurs fit into a world of progress and refinement. No one needed to ask why they went extinct. How could shambling, malformed monsters that looked like a herpetologist's nightmare ever be the pinnacle of life's story? Great catastrophes turned over the makeup of life on Earth, but there was always a sense that the extinct species deserved their fate. That in some way or another, they were simply practice for what was to come.

Experts in the early twentieth century carried on this fatalistic assumption. Dinosaurs were big, bizarre, and anatomically extravagant. The question wasn't why they died out. The real mystery was how they could have persisted for so long, especially when the clearly superior mammals were waiting in the wings to take charge.

Our mammalian conceitedness held on for decades. Even when the disappearance of the dinosaurs became a more legitimate question, the explanations were most always delivered in such a way that the dinosaurs themselves were to blame. The great, trundling reptiles laid eggs and cared little for their young, so mammals feasted on dinosaur omelets. (At this point researchers had paid no attention to the admirable parental oversight of alligators or snakes.) Or dinosaurs clearly invested so much energy and growth into becoming huge, spiky, and strange that they simply ran out of nebulous vital juices. How could a ten-ton rhino look-alike, studded with three horns and a bony collar around its neck, compete with the up-and-coming mammals? The mental capacities of dinosaurs were famously small, to boot. A cold-blooded reptile like a *Stegosaurus* or *Ceratosaurus* was perfectly suited to a lush world of sweltering jungles and dim-witted prey, but the lazy dinosaurs simply did not care to innovate, or even be open to the possibility. And if this is all sounding a little corporate to you, it should come as no surprise that these ideas proliferated during America's great industrialization; "going the way of the dinosaur" is still a phrase used to tar competitors in financial circles.

In time, scientists began to accept the fact that animals do not have internal timers that regulate when species are "born" or "die" according to some cosmic clock, and the ideas about the expenditure of evolutionary energies was misplaced. There had to be some natural explanation. Refining the geological timescale made the question all the more puzzling. Dinosaurs did not represent a primitive lull as the world waited for the rise of mammals. Non-avian dinosaurs persisted for over 150 million years before abruptly disappearing at what seemed to be their apex. There had to be a reason.

Almost everyone had an opinion. Maybe the climate got too hot. Or maybe the climate got too cold. Perhaps some terrible disease ripped through their populations, or sea level rise ruined their favored habitats. Specialists from other fields chimed in, too. An opthalmologist proposed that dinosaurs had terrible cataracts, meaning that the impressive headgear of dinosaurs like the crested shovel-beak *Parasaurolophus* and the many-horned herbivore *Styracosaurus* had evolved as the world's first sunshades. An entomologist spitballed that early caterpillars ate vegetation at such a voracious rate that there was no green food left, meaning that soon after there was no meat, either. Or maybe the time was simply right for mammals. Dinosaur diversity at the end of the Cretaceous seemed low compared with what it had been 10 million years prior. Maybe, after tens of millions of years, mammals started to flex their muscle a little bit and carve out more of the landscape for themselves.

The problem was that many experts focused on dinosaurs alone when the real devastation cut much deeper. Yes, an army of very hungry caterpillars could have denuded Cretaceous forests at a terrible rate, but that explanation did not explain why the flying pterosaurs of the air or the broad, flat rudist clams of the sea went extinct 66 million years ago, much less species of armored amoebas called forams that precisely track the extinction even though their witness testimony to the disaster will never be a cover story. Everyone was so tensely focused on the dinosaurs that the larger pattern was obscured even as experts continued to tabulate the Cretaceous body count.

It was only in the late twentieth century, when the signature of mass extinctions began to coalesce for paleontologists focused on the comings and goings of ancient mollusks and arthropods, that the fate of the dinosaurs started to take on a new gloss. The invertebrate record showed a sharp uptick in extinction at the end of the Cretaceous. The forams and armored balls of algae called coccoliths documented a sudden and horrible event. This is when dinosaurs disappeared, too. Something awful must have happened. Now the question was what.

Experts searched for a compelling cause to explain the devastation. At first, it seemed that some terrestrial trigger was to blame. At the end of the Cretaceous, right when the dinosaur record seems to evaporate in rock strata worldwide, the planet was changing. Sea levels dropped. The climate shifted. Volcanic rifts in the Earth's crust emitted tons upon tons of greenhouse gases into the atmosphere.

It seemed as if dinosaurs simply couldn't keep up with the Red Queen's evolutionary race; they fell behind as mammals kept the adaptive beat. But this story didn't quite fit either. Paleontologists working on the comings and goings of ocean mollusks and other invertebrates didn't see a slow changing of the guard. Better fossil sampling and revised statistical techniques affirmed that life at the end of the Cretaceous was weathering the changes perfectly fine. Then suddenly life suffered a major shock. Something terrible had clearly befallen Earth's biota. The answer didn't come from fossils themselves but from the rock that entombed them.

Battered quartz crystals, vast amounts of prehistoric soot, and a rare metal called iridium, found just at the geological levels where the fossil record of non-avian dinosaurs disappears, suggested that some kind of extraterrestrial body had slammed into our planet. First proposed in 1980, at the fevered height of a new scientific interest in dinosaur biology, the idea set off an academic firestorm akin to the very impact it described. Paleontologists, geologists, and astrophysicists fought as fiercely as tyrannosaurs in conferences and journals over the proper interpretation of the results. But the discovery of an enormous impact crater in the Yucatán Peninsula in the 1990s settled the debate: a massive asteroid about seven miles across had struck Earth at just the moment the extinction becomes apparent in the strata. Nothing like this had ever been seen before. Physicists calculated that the initial impact that created the Chicxulub crater in Central America would have been powerful enough to blow many terrestrial dinosaurs in the vicinity off into space. But it wasn't just the initial hit that sparked the extinction. The aftereffects of this dramatic event tipped the scales against the terrible lizards and many, many other forms of life.

Often, this is about as far as the discussion goes: an immense rock smacked into the planet and myriad species were summarily snuffed out. Simple as that. The asteroid becomes a cosmic bullet shot into the Earth's gut. Yet there have been other impacts of similar or greater scale throughout our planet's history—impacts that did *not* trigger biological disasters. About 35 million years ago, another large asteroid struck ancient Siberia and carved out the Popigai crater, sixty-two miles across. That's more than ten miles wider in diameter than the impact crater in the Yucatán. But this more recent strike did not cause a mass extinction. There was local upheaval and damage, certainly, but life on the rest of the planet kept trotting along much as before. Not all impacts are equal.

The Cretaceous killer thus becomes a special case, standing out from other impacts through time. The size of the K-Pg asteroid, its speed, its angle, and the nature of the rock it struck all came together in the worst possible way for life on Earth—a set of complete happenstances that coalesced into nothing short of an apocalypse. It wasn't just that Earth was hit by a massive asteroid. It's that the aftermath of the impact played out in such a way that life was pushed to the breaking point, with many organisms unable to cope with the rapid changes. Earth swung between a world of fire and ash and one of withering, persistent cold and darkness. Dinosaurs didn't just collapse when the asteroid struck. The real extinction played out over hours, days, months, and years in a constant state of flux as a new world emerged from the cosmic shake-up.

The K-Pg disaster was a global event, its story told through evidence gathered from many places across the planet. But the fossil record is uneven, yielding a collection of pinholes to look through to try to ascertain the whole. As the naturalist Charles Darwin famously observed, the world's geological strata are like a book that lacks entire pages, paragraphs, sentences, and words from the story, leaving us to piece together the narrative from what might seem like isolated parts. By luck, good or bad, some chapters are richer than others. So far as the K-Pg transition goes, the best of these is in the western United States among the Hell Creek Formation beds of central Montana and the Dakotas. This relatively narrow expanse of our planet documents the last days of the dinosaurian reign up through the earliest days of the Paleogene period that followed. The impact boundary is clearly visible in the rock record itself. Sections of these strata displayed in museums look like the world's most regrettable chunk of chocolate cake, dark brown and deadly. In this place we know the cast of characters who ambled across this ancient stage well and can track their fates across time and their changing environment. Their stories tell us how life suffered greatly, yet still survived.

But the reason we've gone back to this place and this one infamous moment is to understand not only why there are no *Ankylosaurus* descendants at the zoo but also how and why we came to exist. The Age of Mammals, a marker literally set down in stone, would never have dawned if this impact hadn't allowed for evolutionary opportunities that were closed for the previous 100 million years. The history of life on Earth was irrevocably changed according to a simple phenomenon called *contingency*. If the asteroid's arrival had been canceled or significantly delayed, or if it had landed on a different place on the planet, what transpired during the millions of years that followed the strike would have unfolded according to an altered script. Perhaps the non-avian dinosaurs would have continued to dominate the planet. Maybe marsupials would have held sway as the most common beasts. Perhaps some other disaster, like massive volcanic eruptions in ancient India that picked up around the same time, would have sparked a different sort of extinction. It's likely that the Age of Reptiles would have marched on unimpeded, but without the origin of any species introspective enough to engage in such ruminations about time and its flow. This day was as critical for us as it was for the dinosaurs.

Now, after decades of fierce scientific debate, our picture of what transpired is starting to become clearer. Paleontologists, geologists, astronomers, physicists, ecologists, and others have assembled a more detailed tableau of what happened to the planet following the collision. It wasn't the impact itself that caused such dramatic damage, but the long-lasting aftereffects that permanently reshaped the nature of life on Earth and allowed for the eventual and unintended emergence of humans. By imagining ourselves in the heyday of the dinosaurs at Hell Creek, on Extinction Eve and what follows, I'm going to walk you through what happened in the seconds, days, months, years, centuries, and millennia after the impact, tracking the sweeping disruptions that overtook this one spot and imagining what might have been happening elsewhere on the globe.

We're about to watch the world change with unprecedented speed and violence. And we've carried plenty of cargo in with us to appreciate the dinosaurian drama. Not in terms of goods and gear, but in our ideas: over two centuries of scientific gleanings that describe everything from how the branching arms of the monkey puzzle tree grow to the taxonomic breakdown of the species that reside in this place. And for all that, the Hell Creek is perhaps the best known of any dinosaurian habitat. It's often both our introduction to the dinosaurs' world and the backdrop for their last great act—overture and finale all in one. This was the close of one era and the beginning of another. As much as we love dinosaurs—enshrining them like scientific relics in our museums, bringing them back to life in film—we know we exist only because they ceded the evolutionary stage to our ancestors. We owe them a debt.

Consider the dinosaur browsing along the forest's edge, sun and shade sliding along its back as the quills growing from its hips and tail lilt with the creature's amble. The name we've given this animal is *Triceratops horridus*, a label codified back in 1889 based upon an even older system of assigning every organism a genus and species name. In our own time, the petrified bones of the dinosaur are incredibly common—known from hundreds of skulls extracted from states huddled around the Rockies—which means we know a bit more about its variations, growth, and behavior than most other dinosaurs, informing our view of the reptilian grazer as its pillar-like legs pump back and forth.

And consider the very idea of the Mesozoic era itself. Geologists and paleontologists have divided this time into three parts—the Triassic, Jurassic, and Cretaceous periods—all delimited by the occurrence of particular rock layers, species, and absolute dating techniques. Dinosaurs appeared toward the middle of the Triassic, about 235 million years ago, flourished in the Jurassic, and kept their influential hold on terrestrial ecosystems through the Cretaceous, spanning over 160 million years as the most charismatic animals on the planet. But from our imaginary vantage point, all these concepts are far distant. The *Triceratops* doesn't know its name, what day of the week it is, or how many millions of years separate us from the tri-horned grazer snuffling around for new growth to munch. Nor does it know of its impending doom.

The story I'm about to tell you, a tale spanning a million years, rests at the nexus of science and speculation. It's a vision of the late Cretaceous and early Paleocene worlds that is informed by everything we've learned through decades of discovery, a tale that emerges from the science rather than a recounting of the scientific process itself. Some of the story is speculative, inferred from hypotheses and available evidence rather than directly referenced in the literature. But much of this tale, the deadly unfolding of devastating ecological change, is based on fact, the scientific skeleton that I've clothed in narrative flesh. This book's appendix explains what's fact and what's hypothesis or invention, and I've divided it out so that we can follow the dramatic changes in the period from 66 million years ago to 65 million years ago without skipping a beat. My goal is to offer an ecological, fleshed-out view of these organisms and their biology during a time of terrible stress, and I've done my best to envision these species as living organisms rather than permineralized, distorted fossils. That's the goal of paleontology, after all—to start with the offerings of death and work back toward life.

I have no doubt that some of my speculations will turn out to be wrong or need modification. Science will continue to make discoveries, refining and enriching what we understand, but I believe we've reached a point where we have a detailed explanation of what transpired at the end of the Cretaceous. I've largely constrained the story to the ancient environs of

Hell Creek because we know that area best, but I've included codas about other places and ecosystems to drive home the depth of this catastrophe. My hope is that fossil discoveries made in the next several decades will give us a more detailed picture of the K-Pg disaster in other parts of the world, far from the haunts of *Tyrannosaurus*. But given the episodic and fragmentary nature of the fossil record, I've often been surprised by how much we've come to know about even just one pocket of the globe during the sweeping changes we're about to witness together.

The change is coming soon. As we watch, listening to the sounds of ceratopsian breaths and the trill of insects, a seven-mile-wide chunk of alien stone gets closer and closer. Somewhere beyond the atmosphere, the end of the Cretaceous looms.

1
Before Impact

66.043 million years ago

The *Triceratops* reeks. Even though only hours have passed since the immense herbivore fell, black clouds of flies buzz around its still nose and glazed eyes—swarms only disturbed by the fluttering and squabbling of the airborne creatures that have settled on the carcass. They are waiting. The food has been laid out, but the opening ceremonies have not yet commenced.

Shortly before his death—a frothing, burning death from the inside, the creep of malignant cancer—the old bull *Triceratops* weighed ten tons. He wasn't the largest *T. horridus*, but he had survived season after season, pushing and shoving other males with his tremendous horns to prove his point when necessary. Every season the competing tri-horns would bellow, cover themselves in shit and mud, and test their strength against each other in great fern-covered clearings that were soon churned into a mess by powerful hoofed feet. But in the past year, the bull started to feel ill. He felt a growing discomfort deep inside. He could no longer compete in the annual territorial fight to mate, and he began to shun the cantankerous company of other *Triceratops*. He was a lone shadow on the landscape, a small, horned hill of muscle and bone who spent much of his time busting rotten logs at the edge of the Cretaceous forest and wallowing in stinking mudholes that helped provide another layer of protection between his scaly skin and the biting insects that had adapted to winnow and nip in the crevices. It was in one such ceratopsian puddle that he one day dozed off and did not rise again.

The pterosaurs had noticed the carcass first. They had it easy. Held aloft on the warm thermals during the day, each of the fuzzy flying reptiles could circle on their membranous wings and spot carrion ripe for the plucking. All they had to do was dip down to pick off a few dried tatters of tendon and putrid viscera. They look somewhat like storks with bat wings, impossible creatures that had nonetheless been the first vertebrates to evolve powered flight. From their insulating coats of primitive down to the hollowness of their bones, this is precisely what pterosaurs evolved to do. Then again, the graceful airborne skills of these soaring scavengers mean little on the ground; while standing, they fold their wings—supported by an incredibly elongated fourth finger—and waddle here and there as if on stilts, a shambling, squawking gaggle. Each trip to the ground has to count. Shuffling about the Cretaceous floodplain expends more energy than simply twirling aloft, making the urge to snag a snack all the more important. And the careless or overconfident do not always make it back into the air.

Nothing of the horned dinosaur's saurian bulk will be wasted. There is no sense of mourning here. A body of this size is food for innumerable bellies, a buffet that will energize the large and small alike as each morsel is taken back up into the ecosystem. A *Triceratops* is no soft piece of meat, though. The pterosaurs and fluttering birds, some with tiny teeth poking from their mouths, have already taken what soft parts they can winnow away from the carcass. Beaks poke at eyes, plunge into nostrils, and tear at the cloaca, the dead dinosaur's body seeming to break out in small sores. But the riches inside require jaw power, a carnivore capable of biting through scaly skin, fascia, and thick muscle to the feast encapsulated in the dinosaur's body wall. For more than a day after the bull expired in his patch of ferns, the flying scavengers have waited. They chitter and squabble and squawk from above and from their perches on the recalcitrantly tough hide of the body, anxious for a meal that cannot begin without the guest of honor.

At last, just as blue and purple light begins to overtake the floodplain on the second evening of the saprophilic watch, an immense shadow steps out from the tree line of bald cypress and ginkgoes. The carnivore is in no rush. Measuring thirty-five feet long, weighing more than eight tons, the grim reptile is at the absolute apex of her ecosystem. Not that she is invulnerable. Fractures, insect bites, and disease affect this creature just as any other. But the only immediate threat to this reptile would be another of her own kind, and even then there are few who are larger than this towering biped—essentially a set of massive, bone-crushing jaws with a body almost entirely dedicated to that purpose. A *Tyrannosaurus rex*, the top predator of Hell Creek.

Her reddish brown hide now draped in orange and gold from the low-angled light of the evening sun, she is absolutely radiant as she approaches the recumbent body. From snout to tail, she is covered in small, pebbly scales with a thin coat of wispy fuzz running from her neck to the end of her tapering tail. A few whisker-like filaments jut from her chin and upper jaws as well, the sensitive flesh all but concealing the ranks of seven-inch fangs in her mouth—an endless supply of fresh, serrated cutting edges that replace themselves throughout her entire life.

Another of her species might take note of her ornaments, too. Her skull is not a smooth sculpture, but a rich topography of hills, valleys, and ridges. A series of bumps sit along the midline of her nose, keratin-covered bosses jutting conspicuously above her eyes to give her perpetual resting dinosaur face. Perhaps these protrusions could be called horns, but their function is not so much defensive as ornamental. Each is a sign of maturity and prowess, easily recognizable to others of her kind and alluring in their own peculiar way during the loud and raucous courtship of the tyrants.

The *rex* keeps her tiny arms close to her chest as she approaches the *Triceratops* carcass. There is nothing for her to grab or flail at here. Each forelimb is powerful, technically capable of holding a carcass in place with two hooked claws on each hand as the terrible jaws go about their work, but most of the time these appendages are a liability rather than a help. An extended arm can easily be broken during the hunt or snapped by a rival in a fight. The entire body plan of the tyrannosaur evolved to emphasize her head, catching and killing and chomping with an overwhelming maw, and so those little arms are largely irrelevant. Still, despite all the weight-saving pockets and air sacs inside her skull, the amount of muscle and bone at the front of her body requires formidable strength behind. Muscular legs suited for long-distance walking and a long, counterbalancing tail keep the *rex* in a near horizontal posture most of the time, although, as some unfortunate mammals and birds learned, she is perfectly capable of rearing back to pluck morsels from the branches of trees as easily as fodder at ground level.

She has gone years without facing a significant threat. She's too big now, secure in her place as an overwhelming carnivore. Much of the mottled camouflage she had when she hatched is gone now, only faintly visible as darker circles along the comparatively light coloring of her throat, belly, and inner thighs. The rest of her body is shaded brownish red—not from dried gore, but because the pattern helps her blend into the shadows at the edge of the forest during her favored hunting hours near sunrise and sunset. After all, a wily *Edmontosaurus* can outrun a *rex* on open ground. Charging directly into a herd of honking duckbills is only a waste of energy and a sure way to send prey snorting and farting their way to the next basin, wary of letting their guard down again. Better to wait near cover, step

step step *lunge,* putting all eight tons behind a devastating first bite that will crush and maim. Even if the quarry manages to survive that first bite, the *rex* can watch and wait until shock and blood loss take their toll. Caution and care are key for this carnivore. Even though the hadrosaurs have no defensive horns like the *Triceratops* and are not covered in tough, crunchy bone armor like the waddling *Ankylosaurus,* those thick, muscular tails can still break ribs. A bucking, kicking *Edmontosaurus* can't kill an adult tyrannosaur outright, but even a fractured arm or shin might lead to infection and pain that could take down even the mightiest predator.

In fact, the worst adversaries of the *rex* are the smallest. The female opens her mouth wide to yawn as she steps closer to the body, a terrible stench rising from her jaws. It is not just the odor of rotten meat and dinosaur breath. Barely visible beneath the great flat tongue are small lesions, the residence of microscopic parasites that are slowly burrowing their way through her jaws. She had inadvertently picked them up from a hadrosaur she had been dining on, a carrier for an organism that relied on both predator and prey for its life cycle. In time, the parasites will proliferate through her throat and leave holes in her lower jaws large enough to stick a thumb into—a set of painful, oozing sores that will make it nearly impossible to eat, much less hunt. But that will not transpire for some time yet. Today, approaching the ripening flesh of the *Triceratops,* she is still the ruler of her patch of Hell Creek.

Her earliest ancestors were not quite so regal. Over 160 million years before, in the early days of the Triassic, the ancestors of dinosaurs were tiny, meek creatures. Covered in downy fuzz, the largest among them stood about four inches high at the hips. They did not rend prey with tooth and claw. They did not have backs plated with armor and spikes. They did not flutter and flap and fly. These small, slender protodinosaurs hopped. The earliest dinosaurs could think of no finer meal than a shiny, crunchy beetle, and they made the most of their hot-running metabolism by skittering and jumping through forests patrolled by much larger, more menacing reptiles.

Even as the protodinosaurs grew larger, about German shepherd size, they still munched on beetles and leaves. They were not terrible, nor did they rule. Ancient relatives of today's crocodiles were the most prominent vertebrates on the landscape back in those Triassic days, in some ways playing out what dinosaurs would later reinvent. Against this background, the very first dinosaurs evolved, and it would take almost to the end of the Triassic for any of them to grow to large size or become prominent parts of the world's ecology. Were it not for a mass extinction, perhaps the earliest dinosaurs would have remained in competition with a menagerie of equally strange reptiles that happened to stake their claim first.

The dinosaurs couldn't have known it, but the humble beginnings of their ancestors made all the difference. The insulating fluff, the hot-running metabolism, the limbs carried directly below the body—all of these traits gave them an evolutionary edge when an intense and exceptional bout of volcanic activity began along what is now North America's East Coast about 135 million years before the age of *T. rex*. This wasn't like some kind of cartoon set to Stravinsky, with gouts of angry lava projected into the air by underground pressure. The Earth oozed and suppurated, lava spreading for miles and miles, releasing tons of carbon dioxide into the air. The seas began to acidify. The climate became erratic, swinging from hot to cold. The hot-blooded dinosaurs survived, as did the warm little protomammals they sometimes dined on, perhaps insulated by their unique physiology and ability to keep warm. But the ruling crocodile relatives were decimated. Some persisted—enough to establish their own varied legacies through the rest of the Mesozoic—but now the roles were reversed. The dinosaurs were supreme, while the surviving crocs scampered and slopped around in swamps; this marked the beginning of tens of millions of years of dinosaurian proliferation.

As for the storied *T. rex,* the tyrants could trace their particular lineage back to the depths of the Jurassic. Their ancestors were small, lithe, three-clawed, and shallow-jawed. They were little nippers, not towering fluff monsters with horrific jaws. Other varieties of carnivorous dinosaurs had beat the tyrannosaurs to the punch—the allosaurs, megalosaurs, and ceratosaurs—and would continue to consume flesh in great volumes for tens of millions of years. Tyrannosaurs were there, but, for the most part, they were small fry—that is, until about 14 million years before the present scene we're witnessing. In the Northern Hemisphere, the terrible predators of previous ages had faltered. There was an opening for the tyrannosaurs. Suddenly, these dinosaurs became predatory giants. The tyrannosaurs became larger and larger, their arms only standing out as tiny evolutionary afterthoughts compared with their huge, muscle-packed skulls. The back of their crania flared to the sides to accommodate even more musculature for slamming jaws shut and controlling prey with powerful neck muscles, so much so that the eye sockets shifted from pointing sideways to staring straight ahead—an unexpected gift of depth perception.

Not only that, but these new tyrannosaurs could smell lunch from the barest odors on the wind. The olfactory bulbs of their brains were larger than the portion devoted to processing tyrannosaur thought. From tooth to tail, the new tyrannosaurs were carnivores unlike any that came before.

In these last Cretaceous days, the tyrannosaurs live up to their title. These terrible dinosaurs control their environments so thoroughly that they suppress the evolution of possible competitors. Smaller carnivores like *Atrociraptor* chase rodents and pluck at carcasses in Hell Creek, and the rare human-sized carnivore like *Dakotaraptor* might frighten a hatchling *T. rex,* but the tyrant lizard king has no real equal in these forests and floodplains. Instead of carnivorous roles being divided between multiple meat-eaters—with one specializing, say, on duck-billed hadrosaurs, another on scavenging, still another on *Triceratops*—*Tyrannosaurus* has taken almost the entire swath for itself.

The dramatic changes the tyrannosaurs go through as they age have allowed them to push out the competition. Changes manifest rapidly as they grow up, with the babies resembling *T. rex* ancestors more than the adult animals. Babies, when they kick their way out of eggs the size of grapefruit, are big-eyed, fuzzy, and leggy, appearing more like toothed roadrunners than nine-ton killers. They snatch insects, lizards, small mammals, and carrion when they can, their skulls remaining long and low as they go through childhood and adolescence. But around the age of eleven, something begins to change. *Tyrannosaurus* youngsters not only keep packing on the pounds, growing at a stupendous rate, but their skulls begin to shift into a different form. Their jaws become deeper and a greater volume of muscle lets them clamp down ever harder on flesh and bone. By the age of twenty, *Tyrannosaurus* can topple large prey and reduce the carcass of a horned or duck-billed dinosaur to little more than splinters and scat. That's why there are no rival species of similar size here. *T. rex* has shouldered them out in a long-term evolutionary arms race in which infant, adolescent, and adult all take up different habits and prey menus, a single monstrous species thriving among the forests and streams of Hell Creek. And they need space. In the whole of North America, there are only about twenty thousand *T. rex*, each with their own territory.

The large female does not keep a lookout for other predators as she nears the carcass, leaving behind huge, three-toed footprints in the soft earth. Even if there were a competitor, she probably wouldn't be deterred. The sweet, rancid stench already wafting from the erstwhile *Triceratops* is intoxicating. She'd picked it up hours ago, just a hint on an early afternoon breeze that shook the magnolia leaves over where she dozed. Through the glades and the towering stands of the dawn redwood, *Metasequoia*, she had followed the aroma of deceased *Triceratops* that grew stronger with each footstep.

She claps her jaws as she approaches. The sharp *thok* sound sends pterosaurs and toothed birds flying, scattering from the body like oversized flies. She wastes no time. Bracing her left foot on the ground, already damp with fluids that have begun to seep from the body, she rears back and draws up one of her great three-taloned feet, raking down the belly. The dead herbivore's thick scales and skin resist the claws for the first stroke, and the second, but the third scrape draws a massive ragged gash as the body wall tears and gives way. A hefty pile of gore slides out, as the great tyrant dips her head down and begins to shovel great gobs of viscera into her jaws.

The *rex* can snap off an entire leg to carry away if she wishes, or pull the head off the shoulders of the dead herbivore. And in time perhaps she will. But there is no need to resort to such violent methods just yet. The softest and most savory parts of the body are still intact. Like a terrible seesaw, she dips her head down and scoops up another hundred pounds of increasingly mashed-up flesh, jerking her head upward to throw the meat toward the back of her throat. Her relatives among the crocodiles and birds can do the same—it is an old ability, called inertial feeding, and it certainly comes in handy for animals that lack grasping hands or the ability to chew their food. Her jaws are essentially a biological set of scissors, best suited to gripping and shearing. The muscles of her gut and the acid in her stomach will have to do much of the work, drawing as much nutrition as possible from the relatively short residence time of each meal. Such is life with a fast metabolism. She can't gorge and then lie in a torpor for days or weeks as every last scrap is liquified. Her predatory prowess comes from her speed and power, and the cost to feed is high. Her body has to act quickly to make the most of each meal before all those delectable tissues are turned into foot-long plops packed with muscle fibers and scraps of bone.

The *rex* feeds greedily, paying little attention to the airborne scavengers that try to nab a scrap or two when she raises her head to swallow another portion. Tasty as the *Triceratops* is, though, the tyrannosaur's eyes are bigger than her stomach.

Almost immediately after swallowing another bolus of meat, she turns and regurgitates all that she has taken in, a pile of deconstructed *Triceratops*. She shakes her great head, snorts, and goes back to the fresh meat as the pterosaurs and birds quickly engulf carrion à la *rex*. There is enough for all the ravenous mouths.

Hour by hour, the carcass becomes smaller. Bones that were scraped with peglike teeth to access the thick muscles are now pulled and smashed in great jaws to get whatever scraps and marrow might remain. The female presides over the proceedings, and she is not an ungracious host. There is only so much she can eat at a time, after all, and when she goes off to groggily doze in the shade of the nearby palms other carnivores carve out their own shares. In singles and in pairs, small dromaeosaurs pick over the raggedy bones—a table scrap to the *rex* is a mouthful to them. Still, some chatter and argue over who could claim rights to the bounty, leaving brown and white feathers strewn around the bones as each bird, raptor, and pterosaur tries to get a little something for themselves. The dance goes back and forth, the *rex* taking her share and the hangers-on picking at the rest, order within the grisly chaos of the scene.

Soon there's little left but a reeking three-horned skull covered by a jerky-like coating of dried-out and exsanguinated flesh. All the best cuts from the limbs, the hips, the tail, and the neck have been taken. The *rex*, her belly now full of more bone than meat, snuffles at the battered skull and gives it a push with her nose. There is nothing left for her here. The scent is no longer that of food but of death, and she leaves the remnants to whatever scavengers might have it—the flies, the burrowing beetles that will raise their young in the bones, the fungi that will grow out of the enriched earth, and the uncountable microorganisms that will complete the great recycling project. This *Triceratops* will not join its relatives in the fossil record. Only the trampled earth testifies to the fact that the herbivore was ever there.

Sated, the *rex* begins to wander again. The only nest she knows is the one she makes each night within her territory, a patch of shifting boundaries that requires constant marking and defense against others of her kind. Still, her reptilian brain carries an expectation that she will not see another tyrant. If she does wander into a clearing and sees another *rex*, or spots a great fuzzy shape slinking through the trees, she will meet the threat to her kingdom with a full-throated challenge of her own.

But all is calm on this morning. The sun is still low, casting golden light over the marshes and stands of trees. The *rex* continues her aimless patrol, her senses idly scanning for anything of interest as she puts one three-toed foot in front of the other. Even if she doesn't have to hunt now, the need will soon arise again and it is better to be situated in a place where game is likely to come by. Not just any prey will do. The Hell Creek ecosystem is pulsating with life. Insects stridulate and sing from the trees. Crocodiles and odd croc-like copycats called champsosaurs snap at fish and small turtles in the weed-choked swamps. Fuzzy mammals chitter at each other from their burrows and tree branches. Lizards scamper over tree trunks and snakes slither through the fields of ferns. Every morning begins with a chorus of birds and pterosaurs all talking to members of their own species, inadvertently making a riot of sound that can wake most any dinosaur from its slumber. The *rex* can technically make a meal of any of these, but smaller fare is usually reserved for desperate times. Her favored prey has to be worth the effort, large and unwary enough to be caught by surprise.

Now and then she will munch on some of the midsized dinosaurs, especially when she's lucky enough to stumble upon a carcass. But most are either too quick or not worth the trouble. The ostrichlike *Ornithomimus* are easy to spot, especially with their elaborate arm feathers, but they run much too fast. The dome-headed *Pachycephalosaurus* are also quite nimble and, in any case, are not found in great numbers. The smaller raptors often flee when the tyrannosaurs approach and, in her experience, only get within biting range when they try to feed from the same carcass. And while the beaked, bipedal *Thescelosaurus* don't have any defenses besides their natural camouflage, the herbivores are barely worth the energy needed to grab them. For a *rex* of her age and size, going meal to meal is exhausting. Better to go all in on a buffet than try to stuff herself on hors d'oeuvres.

Of the three large herbivores common to the area, only two are worth hunting. The great shovel-beaked *Edmontosaurus* walk together in herds. All those alert eyes test the tyrannosaur's ability to sneak, but, then again, the false sense of security makes for a greater number of comfortably unwary targets. Not to mention that the broad, thick tails of these dinosaurs supply enough meat for a couple days just by themselves. *Triceratops* are a little trickier. These three-horned herbivores group together as teenagers but eventually split up as adults. Sometimes they gather together to graze, but they don't form organized herds. All the same, taking down a *Triceratops* requires both practice and luck. Their long horns can inflict terrible wounds and infection; they must be attacked from behind with a first horrific bite before they can wheel around and face their enemy. But even that is better than dealing with *Ankylosaurus*. The low-slung herbivore is not invulnerable, but these plant-eaters have armor down to their eyelids. A tyrannosaur might lose several teeth just trying to find the right place to bite, all the while risking a whack from a tail club the size of a car tire that can easily shatter bones. A *rex* might be able to stave off a greenstick fracture from a struggling *Edmontosaurus* or a slash from a *Triceratops* horn, but a *thwack* from an *Ankylosaurus* club can be devastating, even deadly. The only time the female had dined on an ankylosaur, rolling it over and digging into the softer, gas-distended belly, was when she happened across one that had died of natural causes.

None of her favored prey seem to be nearby on this particular morning. Birds take off from the undergrowth at her approach, and mammals bark their displeasure with their fuzzy false bravado, but she has to go some way yet to find a suitable hunting ground. And she has become thirsty. As the midmorning sun begins to erase the long shadows of morning, the *rex* steps through a stand of dawn redwoods growing by a small pond and dips her head down to drink. A turtle slips off the log it was reclining on and into the water, its retreating shell quickly obscured by a cloud of bottom muck. Then another *thunk* as a toothed bird waiting on a branch plummets below the surface, rising just as quickly with something small and silver in its beak. The *rex* watches the bird as it flies off, not hungry so much as curious. Everything often goes so still in her presence that motion always draws her eye. With a snort, she tilts her head back down for one more great, sloppy mouthful before shaking from snout to tail and turning back into the ranks of trees.

She will not feel the violent, vital pulsing of fresh blood in her mouth again. Unknown to her, and to all the organisms of Earth, the end of the world is speeding ever closer at the rate of twenty kilometers a second.

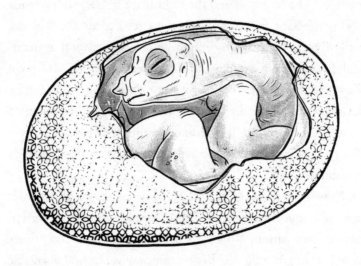

Elsewhere
NORTH HORN FORMATION, EASTERN UTAH

The tiny dinosaur kicks and turns inside its own private pond, grinding its almost microscopic teeth as the embryo settles into a more comfortable position. It won't be long now. The confines of the hard-shelled egg are becoming increasingly cramped by the long-necked creature's rapidly expanding size, and the pointed, temporary egg tooth needed to poke that first critical hole in the shell already juts from the embryo's snout like a ridiculous prosthesis. Over the past several months the baby has gone from a blob of rapidly dividing cells to a creature recognizable as a dinosaur. Nerves, blood, muscle, bones, teeth, and vital organs have all coalesced into an animal that paleontologists will one day name *Alamosaurus sanjuanensis*— one of the largest animals to ever walk the Earth.

The nascent titan, one of a clutch of a dozen eggs, was plopped into a shallow sandy nest by a mother over eighty feet in length and weighing more than twenty tons. Some adults grow larger still. The biggest stretch more than a hundred feet from the front of their pencil-toothed muzzles to the tips of their tapering tails. At the shoulders they stand taller than a *T. rex* on tiptoe. Not that the tyrants could ever get close enough for such a direct comparison. A swipe from a muscular tail or a stomp from a pillar-like leg can irreparably batter and break even the largest of the fuzzy carnivores. At less than half the length and a third of the weight of *Alamosaurus*, tyranno-saurs usually do not prey on these giant herbivores, but wait until one dies from some other cause. Something as simple as a bacterial infection could provide weeks of food to a *T. rex;* the carnivorous gourmand could settle inside the titanosaur's body and just eat its way out.

Not that all tyrants get to enjoy such feasts. *Alamosaurus* is a migrant from the southern continents, one of the bulky, thick-bodied titanosaurs that have proliferated through Cre-taceous Africa and South America. Despite sharing the same basic body plans as earlier sauropods like *Brontosaurus*, the ti-tanosaurs are much burlier. An *Alamosaurus* and a *Brontosaurus* of the same length would be dramatically different in weight, so much so that even their footprints document their disparity. The giants of the Jurassic, like *Brontosaurus* and *Diplodocus*, carried most of their weight over their hips. Their broad hind feet pressed into the sediment deeper than their front feet. But

Alamosaurus is a front-wheel-drive dinosaur. Titanosaurs carry the center of their mass closer to the shoulders, their arms wrapped in powerful muscle, and therefore leave tracks with deeper front foot impressions than other long-necks.

Such giants are the dinosaurs who eventually made their way back north toward the close of the Cretaceous, revitalizing a niche that had remained abandoned for tens of millions of years. Duck-billed dinosaurs, horned dinosaurs, and ankylosaurs are the big herbivores of the northern continents, their ability to shear and crush and chew through vegetation as fibrous as rotten logs giving them an evolutionary edge. Sauropod dinosaurs were missing from the picture, and the arrival of *Alamosaurus* in Cretaceous North America marked the return of giants who scarfed down tons of vegetation to be fermented and broken down inside their vat-like guts. The methane reek can get overpowering when the giants gather.

Alamosaurus didn't spread everywhere through the northern continent, though. The plant-hungry species has never made it as far north as Hell Creek. Prehistoric New Mexico, Texas, and Utah are about as far as this enormous sauropod has reached, just within the southern range of the ultimate tyrannosaur. The North Horn environment is one of the only places in the world, in any era, where bone-crushing tyrannosaurs can chomp on a sauropod dinosaur.

To such predators, the gestating *Alamosaurus* is less than a mouthful. The egg enclosing the baby is only about the size of a softball. The infant, after breaking out of the egg, will be no bigger than a house cat—nothing more than Cretaceous popcorn to even a teenage *T. rex*. Parent alamosaurs do little to discourage such predation, unlike some other dinosaurs, who watch over and protect their young, sometimes for years. Like other sauropods, however, *Alamosaurus* simply lay their eggs in large clutches, watch over the nests as the babies incubate out of ancient instinct, and depart near the hatching season. The infants are on their own, their peeps heard only by each other—and the small mammals, crocodiles, snakes, and feathery dinosaurs who feed on them—when they break out of their nests and flood the Cretaceous forest. This is why each skeletal element of their legs, which initially form as cartilage, turn to bone and fuse before the rest of their diminutive skeletons. They have to be ready to run from the second their little paws touch the outside soil. From there, luck will decide who survives the critical first year. Most will not.

2

Impact

Itch. The annoying impulse repeats itself over and over again through the young hadrosaur's body. *Itch itch itch,* a terrible tingle between the dinosaur's toes and along his scaly flanks.

There's only one thing for the young *Edmontosaurus* to do. Sheltered by the shade of the forest, on the edge of the dinosaur's small herd, the shovel-beaked herbivore stops to relieve the tingling burn. A beech tree will suit the situation nicely. The snarl of trunks sprouting low to the ground creates a series of great riblike scratching posts, each wrapped in sufficiently coarse bark. Tilting his thick, hefty tail backward to seesaw the front half of his body up, the dinosaur slowly steps and scrapes and rubs against the rough trunks, the friction sending momentary relief over the pebble-like scales covering his body.

This particular *Edmontosaurus*, still a young male at about eighteen feet long, is not the first to visit this particular post. The bark of the beech and the surrounding trees—sycamore, dogwood, laurel—have been polished, rubbed, and broken open by dinosaurs desperate for some form of relief. Gooey sap has oozed here and there, collecting small flies and other unlucky invertebrates before hardening into resin. Someday, perhaps, these globs will be amber. For now, they're just a mess and a mark left by the megaherbivores who call this part of the floodplain home. Generations of dinosaurs have marked their patch of forest in just this way, each scratch prompting the same behavior until the herds of *Edmontosaurus* know their own favorite spots.

The juvenile lets out a satisfied honk as he goes back down on all fours again, mitten-like gloves of flesh around his hands touching the ground as the great, three-toed hind feet start to move the dinosaur along again. Satisfying as the scratch was, lagging behind for too long is a risky move. While the *Edmontosaurus* is large enough to avoid the jaws of the smaller predators like *Acheroraptor*—and perhaps even ward them off with a great swing of his muscular tail—he is still small enough to be a day's supply of food for a large *T. rex*. Best not to tempt a sudden bite to the rump. Step by step, the *Edmontosaurus* catches up with the rest of the herd and wiggles his way back into the crush of dinosaurian bodies as the duckbills trample their way through the woodland.

The healthy dread of sharp teeth and watching eyes in the

forest shadows has led *Edmontosaurus* to unintentionally re-shape the environments around them. Theirs is a landscape of fear, and the movement of the herd outlines the boundaries of this invisible, shifting territory.

A lone *Edmontosaurus* does not have any extravagant defenses. These dinosaurs lack the horns of *Triceratops* or the hard scutes embedded in the skin of the *Ankylosaurus*. A large adult can kick, swing its tail, and flail about if attacked, but by then the damage might already be too catastrophic to survive. By adulthood, every *Edmontosaurus* has seen this happen—a sudden lunge from the edge of the forest, a blur of great jaws erupting from the shadows to crush and consume.

And so *Edmontosaurus* have become wary of the forest's edge. The scores of tree trunks can conceal a stalking tyrannosaur all too easily, with perhaps only the snap of a branch or the screech of an alarmed bird to provide any warning. Instead, the open meadows and floodplains feel more comfortable for *Edmontosaurus* herds. Here, there are no surprises. Any stalking *T. rex* will have to step out of the shadows to be seen, by which time the hadrosaurs will have ample time to use their one advantage in such a face-off—speed.

For all their terrible chomping power, adult tyrannosaurs can't move very fast. Most of the time they amble along at a slow stroll. In a rush, they can run at speeds up to fifteen miles per hour. But they don't need to move fast. They did not evolve to chase down prey. Instead *T. rex*—like the other great tyrannosaurs before it—evolved to hunt by ambush, waiting until just the right moment to put the dinosaur's leg power into one great lunge. If the tyrannosaur misses and the prey takes flight, or if the tyrannosaur is spotted before this critical moment, the hunt ends and the great carnivore moves off. And over the millions of years that tyrannosaurs and hadrosaurs have coexisted, this hunting strategy has molded the attributes of *Edmontosaurus*. Given room to move, adult edmontosaurs can rear up on their hind legs and take off at about twenty-eight miles per hour. That's far from being a dinosaurian land speed record, but it's enough. If one *Edmontosaurus* in a herd spots a tyrannosaur and honks in alarm, the whole herd stands upright and bolts, leaving the tyrannosaur to try to dine another day.

Naturally, *Edmontosaurus* are not on their toes at all times. Some of them do get caught, especially when having to cross through patches of forest to get at their preferred grazing spots. Much like carnivores of any era, the tyrannosaurs intentionally target the old, the sick, and the young, largely leaving healthy adults alone. While there's nothing on an *Edmontosaurus* that can impale or bash a tyrannosaur, hadrosaurs can still deliver powerful kicks and thrashes that can break bone. So *T. rex* often tries to pick out younger animals when they can, creating generation after generation of evolutionary pressure to grow up faster. By growing rapidly, *Edmontosaurus* race toward a body size that's less likely to be targeted by carnivores. For the most part, if edmontosaurs can survive their first year their chances of survival steadily increase until they reach old age and their joints begin to creak with arthritis.

The nature of the Hell Creek landscape is pruned and shaped by these interactions. While it's certainly true that plants form the foundation of any food web, ecosystems are also affected by top-down forces between the animals. An ecosystem does not simply grow from the ground up but is somewhat squeezed between various pressures—landscapes evolve just as animals do. And so the *Edmontosaurus* herds are unknowingly helping to ensure their own survival by maintaining open fields of low-lying vegetation.

Without the *Edmontosaurus* or *Triceratops,* much of Hell Creek would likely be forest. Young trees would compete with each other for space and sunlight, but the Cretaceous woods would grow far thicker with a greater number of saplings allowed to seek their dendrological potential. With such ever-hungry megaherbivores around, however, the grazers pluck up many of the low-lying plants and young trees before the shoots and saplings have much of a chance to become established. The *Edmontosaurus* herds are effectively weeding a Cretaceous garden, maintaining the open spaces that they prefer. More than that, the repeated movements of these dinosaurs create game trails and gradually depress the soil. Season by season, divots in the dirt have become puddles, which have become ponds, altering the habitats through little more than force of dinosaurian habit.

All the same, even idyllic habitats carry dangers and annoyances. *T. rex* isn't the only creature in Hell Creek that feeds on dinosaurs. In fact, the young male edmontosaur is already being eaten. The itch is a feeling, a sensation to be relieved, but the herbivore has no idea what causes it—lice.

An *Edmontosaurus* is a cozy home for a louse and its prolific, extended family—foot after foot of supple dinosaur flesh just waiting for tiny piercing mouthparts. While the dinosaur's skin is tough overall, and very good for avoiding abrasions and scratches and bruises that come from wandering through forests, the great swaths of skin also create folds and indentations where the limbs meet the body and along the neck. These hot, dark places are comfy for lice and other biting arthropods, single-minded parasites nibbling away at the dinosaur dermis until they break the skin open *just enough* to have a capillary offer up some blood. All those blood meals fuel the next generation, which continue in the same tradition as their parents. And given the bumping, thumping, humping, and other forms of skin-to-skin contact among the edmontosaurs as they move along, the lice have plenty of opportunities to spread. To the adults, the lice are primarily annoyances. The insects itch and create sores, but they are as much a fact of life as the sun overhead or the breeze that wafts through the magnolia blooms. But hatchlings might be overrun by lice if their parent carries a heavy load, and all the effort required by the immune system to fight off the bugs can stunt the growth of the younger members of the herd. The best the hadrosaurs can hope for is trees to scratch on, mud to roll in, or a night that is cold enough to kill off some of the lice.

The herd has moved out into the open now, leaving the sheltering tangle of forest as they enter a broad clearing dotted with ferns and cycads, interrupted here and there by palms jutting straight into the air. This is a saturated, turned-over place, once a great green lunchbox but now a stopover snack station that has yet to fully reseed. Edmontosaurs and *Triceratops* have changed the landscape itself. The herbivores eat seedlings, push over trees, and churn the soil with their steps, keeping the Cretaceous meadow sparser than it used to be. The hungry dinosaurs munch anyway, lowering their long, square-muzzled mouths to the ground to nip and crunch at the vegetation. Despite the name they'd take on millions of years in the future, these are no duckbills. They are more like shovel cows, beak and teeth working together to break down even the coarsest vegetation. The dinosaurs' long snouts are tipped in a square, ridged beak that juts downward—a shape adapted to plucking up lots of low-lying vegetation quickly. And, unlike the actual ducks that would eventually evolve among the avian dinosaurs, every *Edmontosaurus* has an impressive battery of more than a thousand teeth crammed together. Individually, each fresh tooth looks roughly diamond-shaped. But put together and worn down by plants that are packed with dirt and have themselves begun to evolve defenses against herbivores, the teeth make great flat grinders that work like massive molars—pulverizing and pulping whatever is placed into those cheeks. In cooler months, a hadrosaur might chaw rotten logs to get their daily fiber—with mushrooms and insects adding a little protein to the mix. But the teeth of these hungry hungry herbivores are only part of their impressive oral apparatus. The skull bones of these dinosaurs have give and flex to them. They are not locked in place as if the whole skull is a big set of bony scissors. As the lower jaw comes up with each chewing stroke, the bones holding the upper teeth are flexed out and to the side, smearing and grinding plant matter against the teeth. These teeth go back to their resting position when the jaw opens again, ready for the next round of crunching. It's an incredibly impressive anatomical setup, the majesty of which is somewhat offset by the fact that every *Edmontosaurus* has a small, thin dome of flesh on top of its head to help herd members know who's who. It's a ridiculous ornament, but one that these dinosaurs seem to find quite fetching.

The herd is calm as they snort and graze. There is no whiff of carnivore on the air, no stink of carrion to make them grow wary. The day is bright. A soft breeze momentarily mitigates some of the day's heat. The dinosaurs are left to munch undisturbed. Each and every one of them, all thirteen members of the small herd, is totally unaware of what just happened.

There was no impending sense of doom. There was no shift to the wind, or darkening of the clouds. No lightning, no thunder. In this little patch of Hell Creek, Montana, all is as it ever was as far as the dinosaurs are concerned. But more than two thousand miles away, a chunk of extraterrestrial stone more than seven miles across just slammed into the Earth. This is how the end of the world starts.

The rock did not simply appear out of nowhere, of course. The deadly stone has its own history, an origin that has unfolded over the millions of years that dinosaurs were living and dying, evolving and going extinct. The makings of this moment began long ago and millions of miles away, chance events that stacked one atop the other with deadly happenstance and can only be understood in retrospect—the deduction of devastation. The unraveling of life on Earth began in cold, dark, lifeless space.

The Oort cloud is an ancient haze around our solar system. The number of objects within it is incalculable, so large that numbers start to lose meaning. Millions of objects, billions, trillions. Even if you were there at the moment when the rock first formed, you could spend your entire life counting all the icy chunks of stone and debris here and die of old age before you made much of a dent in the number. There is a stark beauty here, but it is also our solar system's scrapyard—a place where meteors, comets, and asteroids are born from remnants of rock that were never aggregated into a planetary home.

Within this immense field of chilly debris, there is one rock relevant to our story. It is a solitary and solid chunk of stone whose cold indifference will, in time, make it the deadliest object life on Earth has ever encountered. It's huge, a stony leviathan drifting through space. Its rocky composition makes it a carbonaceous chondrite, an ancient form of space rock filled with tiny dots. These specks, called chondrules, are molten drops of minerals like pyroxene that became wrapped up into the stone. The rock is a remnant of a much more ancient time that has somehow avoided incorporation or obliteration.

Precisely where this deadly crag came from is lost to time. Perhaps it was once part of a planet that was knocked off by another impact. Maybe it is a leftover piece of planetary formation, the interstellar equivalent of leftover dough sticking to the side of a mixing bowl. It's been traveling along for some time, part of an immense igneous school, an unsettled mix of suspended rock that responds to the flow of the universe. It has struck other rocks, and other rocks have struck it, giving the pockmarked appearance of a harsh adolescence in space. Tens of miles across, it's too large to be totally destroyed and so it keeps picking up scars.

But something has changed. An invisible, irresistible pull has begun drawing the stone toward the center of the solar system. This isn't quite like a tractor beam in science fiction, but the idea still suffices. The immensities of the ancient sun and Jupiter are too much to avoid as the asteroid makes its circumnavigation of space. Gravity draws the rock nearer. The sun is 333,000 times the mass of Earth. Jupiter is a paltry 333 times as massive as Earth, by comparison, but it's still enough. Between the two of them, there is enough pull to shunt massive rocks onto new paths through the solar system. Over time, bit by bit, the stone draws nearer and nearer the 470 million miles between the sun and Jupiter.

The span from the sun to the planet is so great that it would seem like an asteroid could just disappear in the space between them. The void in the middle is more than two and a half times the distance between Earth and Mars. It's an open space with millions of miles in any direction, a gap that makes even planets seem small. The rock might easily find its place there and take up a new residence, rotating as all those little impacts wear it down to mere components of its original self. But that's not what is going to happen. Drawing ever nearer, the asteroid begins to thaw. And with heat comes expansion. Weak spots begin to form through the massive chunk of stone as it gets ever closer to Jupiter's surface. Inch by indiscernible inch, it seems set on a collision course with the planet.

The gravity starts to be too much. The asteroid is pulled, stretched like cosmic taffy, changing from a subcircular lump into a long cylinder. This is the tension before the snap, the asteroid stretched like the rubber of a slingshot before the release. And then, *pop*, it breaks. A stone that withstood knock after knock during its long limbo on the edge of the solar system is suddenly broken into smaller chunks. A few of them crash into Jupiter, plummeting beneath the gas clouds to the surface below. But not all. There is one piece in particular, more than seven miles across, that escapes the gravitational invitation and sails out from the solar system's center toward a more distant region. Given a little boost from its close call, the rock starts moving at impressive speed toward the third planet from the sun.

This is an entirely accidental extinction skillshot. Other asteroids and comets have been made the same way over time, but most have missed Earth. They've sailed clear by various margins, with the evolving organisms of Earth none the wiser. Even if the denizens of Earth knew about what is now heading their way, there is nothing to be done about it. This time, the great rock is going to hit. It's not going to get bumped off course by another asteroid. It's not going to burrow into Mars and crack the Red Planet's dry surface. It's not going to slam into the orbiting moon, as many other rocks have, making lunar seas and craters. Out of millions of potentially deadly rocks, this is the one. This is the accident that will exact an awful toll on Earth's species, but without malice or vengeance. It's both end and beginning, a period that will punctuate the Earth and create a stark dividing line between the seemingly endless Age of Reptiles and the fiery dawn of the Age of Mammals.

The rock is moving fast. So fast that if we were to stand at a single point and try to watch its passage, we would feel it rather than see it. The asteroid is hurtling through space at a speed of 44,738 miles per hour. There is no measure for comparison. Few things in our experience flash through the universe at such speeds. It's not even "blink and you'll miss it" fast. It's "watch for it and you'll miss it" fast.

Earth itself has a role to play in this pivotal moment as well. Our planet does not sit stolidly in space; it turns in its orbit, moving at 67,000 miles per hour around the sun. That's even faster than the hurtling asteroid. All of that celestial speed goes unnoticed until two traveling bodies come into catastrophic contact.

For a brief, terrible moment, the asteroid draws a bright line across the sky. The friction of the great rock entering Earth's atmosphere, the air itself rubbing against the stone to make it glow and break off some of the outer pieces. But this is no pebble that will burn up before striking the ground. It's miles of rock, and there is no way for the planet to dodge it. The stone hits home in the ancient Yucatán Peninsula, too fast to see, at a deadly angle of 45 degrees, a shot straight to the center mass of the Earth.

The whole of Hell Creek takes no notice of this moment. Feeling mostly sated, the *Edmontosaurus* herd begins to shuffle on in search of suitable shade to while away the rest of the afternoon and perhaps even find a place to take a soothing mud bath somewhere along the coastal plain.

The herd passes a young female *Torosaurus*, a three-horned loner rarely seen in this part of the continent. About twenty feet long from the tip of her beak to the end of her tail, she is still small for her species. Fully grown, the herbivore will tack on another ten feet and weigh as much as six tons—certainly enough of a match for any hungry tyrannosaur that might think of making her a meal. But, for now, she is only just beginning her life as an adult. A different itch from those suffered by the hadrosaurs, a biological nudge to mate, has left her feeling antsy. She scrapes the keratin-covered horns over her eyes against the bark of a dogwood tree, the fresh wounds oozing sap down the gray trunks. Soon she would seek out a mate.

From the bones alone, it's difficult to tell the difference between female and male *Torosaurus*. Their skeletons are the same size, with the same pattern of ornamentation, and the same great horns jutting from their face. All of these signals are part of how these herbivores recognize each other at a distance. *Triceratops*—a close evolutionary cousin—live in greater numbers here, and they look very similar. The principal differences are in the great bony frill jutting out from the back of their skulls. In *Triceratops*, the great neck shield is solid bone through and through. In *Torosaurus*, the outside of the frill is dotted with double the number of little triangular bones—called *epiossifications*—and two great holes sit on either side of the frill's midline, making the great crest a little lighter to heft around. The ornament has little to do with survival. A cunning *T. rex* can outstep either ceratopsid and sink its teeth into the vulnerable neck behind. But, for *Torosaurus*, recognizing the frill is still important. In her species, females show off and males watch, waiting to be chosen as the females push back and forth to secure mating territory. Try to court a *Triceratops* and she might find herself gored for the trouble, her advances taken as aggression. And with so few of her species in this part of the continent, she needs that signal to be sure.

She has some advertising of her own. For almost her entire life, the scaly skin over her frill was shaded light brown with lighter cream dots where the skin stretched over the frill apertures. Now, however, the middle of those shallow cream depressions is beginning to blush red, the blood vessels following some ancient sense of timing to change the frill's color. The frills of *Triceratops* do not do this. Those keratin-covered ornaments don't change at all. *Torosaurus*, however, blushes.

Navigating the intricacies of *Torosaurus* courtship is another matter. Physiology can only take the young female so far. She has seen other males and females bellow, mock-charge, and display to each other, turning broadside to show off their size or strength. Females always make the final call, either accepting the male or running them off with pokes of those three-foot prods sticking out from over their eyes. But that was all she knows. Like most non-avian dinosaurs, *Torosaurus* reach mating age before reaching their adult size. She is still growing—and fast. She won't reach her full skeletal stature for another decade. Her inheritance from her distant, early ancestors, though, is a reproductive tempo that runs quickly. The ruling reptiles outpaced early mammals by producing large numbers of offspring and having a life cycle that includes mating before reaching full maturity. These simple quirks of biology are what allowed the Age of Reptiles to begin so many millions of years earlier, a flood of saurians spreading through ancient habitats and remodeling them as the reptiles evolved. And so it remains on the last day, an imperative to create a future that will never arrive.

To call the Hell Creek ecosystem the pinnacle of dinosaurian evolution isn't quite right. The landscape is full to the brim with dinosaurs, but that has been true of other ecosystems in other places and times. Over 80 million years earlier, in this same part of the world, multiple species of enormous, long-necked herbivores wandered across fern-covered floodplains, while carnivores similar in size to *T. rex* stalked through tall conifer forests. This Late Jurassic world was the first time dinosaurs cast truly long shadows over Earth, attaining unprecedented sizes and forms. The smallest were pigeon-sized, and the largest were more than a hundred feet long from the tip of their pencil-toothed muzzles to the end of their whiplash tails.

Dinosaurs did not exist separately from the rest of the landscape. The entire reason that such stupendous organisms were able to evolve in the first place rested on so many other changes to the vegetation, insects, mammals, and other organisms they shared their world with. The world was not just made up of plants to create food for herbivores that were in turn fed upon by carnivores. Each species has had multiple connections and roles, generating ever more diversity as evolution continued to roll on generation by generation.

The Age of Dinosaurs has been a great blooming of life. In some respects, it's strange that it's gone on for as long as it has. Mass extinctions are random, not cyclical—there is no pattern or timetable to their occurrence. From the time that early animals began to flourish in the seas during the Cambrian, it seemed that life couldn't go more than 90 million years without some sort of dramatic die-off. Sometimes it was even less: only 50 million years separated the worst mass extinction of all time and the Triassic disaster that allowed dinosaurs to stage their ecological coup. But from that time onward, there has not been another mass extinction. For 135 million years, life has thrived without the threat of decimation.

Diversity tends to generate even more diversity. Perhaps that seems counterintuitive given what's about to transpire and the narrative that often frames the aftermath of impact. Mass extinctions are terrible events that inadvertently clear the field for the survivors. The entire ecological edifice of interconnected species is torn down, leaving the remaining species to plow new niches. There is no evolutionary directive that indicates any particular species must exist. There was never any adaptive gap that required a *T. rex* or a *Triceratops* to come into being. Such creatures evolve through the push and pull of variation and natural selection, a never-ending dance that changes the partners in the process. Mass extinction events are destructive, but they do not necessarily spur new growth by themselves. In fact, some of the grandest moments in evolutionary history—when life truly burst forth and filled the world in ways never before seen—were not at all connected to mass extinctions. Nature's greatest flowerings have often come from evolutionary accidents and happenstances dealing with organisms themselves, such as the ancient time when fish with legs began to crawl out onto soaking mudflats and plowed a life at the water's edge. Life was no longer confined to water by this time, but vertebrates munched on insects and plants that had already made their homes on land, changing to meet a new environment and eating new prey, leading to a burst of adaptation.

And so it was during the heyday of the dinosaurs, through all those millions of years from the Jurassic through the Cretaceous. Giant herbivores like *Alamosaurus* didn't evolve because such creatures were in any way predestined or even necessarily needed. Dinosaur giants arose thanks to the push and pull of different evolutionary forces. Growing large quickly was a defense against carnivores that were themselves evolving to become larger in a constant evolutionary arms race. Huge saurians were able to evolve at all because large size comes with a physiological profile that allows big animals to subsist on vast quantities of low-quality food, supplied by quick-growing vegetation like ferns. And as these dinosaurs grew larger, nudged toward fast growth and titanic sizes, gaps opened up for smaller species to run around the same habitats, not to mention all the creatures—such as parasites and dung-eating insects—that benefitted from having lots of dinosaurian real estate to dig into. Life itself generated the circumstances for new species to evolve, variation making room for even more variation just as trees that grow tall in a forest leave room for distinct environments in their canopies, their understories, and on their trunks. This vibrance takes time to build, and the Mesozoic was Earth's equivalent of an endless summer where reptiles lived large.

Mammals have enjoyed their own heyday during this time. Perhaps that sounds strange. Traditionally, mammals are cast as evolution's underdogs, tiny beasts that squeak out a living eating bugs during the twilight hours when dinosaurs begin to doze. The stereotypical image is of a tiny shrew gnawing on a prehistoric cricket. But the earliest mammals evolved about the same time as the first dinosaurs, and they've been thriving alongside the saurians. True, they've never gotten larger than the size of a house cat in all these millions of years, but they are not all timid insectivores. In the shadow of the dinosaurs there have been mammals that look and behave like prehistoric beavers, flying squirrels, aardvarks, raccoons, and more. They did not evolve because dinosaurs ceded any ecological space for them to arise. To the contrary, mammals evolved to take advantage of ecological opportunities at a small scale. The evolution of social insects like termites, for example, provided a food source for mammals capable of digging into soil and rotten logs. The sheer number of baby dinosaurs on the landscape each season led to the evolution of predatory mammals that ate eggs and reptilian infants. Tall forests, with canopies too high for even the biggest predatory dinosaurs to reach, provide refuge for mammals that hop from tree to tree on parachutes made of furry skin. Perhaps mammals have not literally shaped the landscape in the same way as non-avian dinosaurs, but they are not simply waiting for their deliverance. They, too, have changed the world simply by existing in it, and those alterations have only generated more diversity through age after age.

The proliferation of Mesozoic mammals is critical to what's about to transpire. Had there been few mammals, or if mammals had all been slight variations on the theme of insectivore, the beasts would be incredibly vulnerable to extinction. Variety is not just the spice of life. The seeds of survival are sown through diversity. In a time like this, when the odds are going to turn against all, even slight differences can settle who survives and who perishes. The greater the diversity, the greater the chances that some species will be able to persist.

Hell Creek seems to have as many mammal species as it does dinosaurs. They require a frame shift to see, a focus on life at a smaller scale. A low head with a wet nose cutting a V across a stream is an opossum-sized *Didelphodon,* an omnivore that munches on everything from snails to little dinosaurs. High up on the branch of a dawn redwood, a ratlike mammal with a long, prehensile tail—*Cimolestes*—searches for bugs hidden in the bark. Down on the ground, the pointed nose of a squirrel-like mammal called *Mesodma* snuffles just at the surface, checking for scent before emerging from his burrow. In mere moments, these beasts will face the same terrors as the dinosaurs—the greatest test to life's profound variation in over 100 million years.

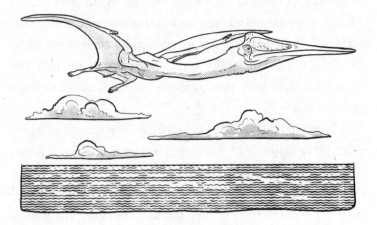

Elsewhere

SOMEWHERE OVER THE NORTH ATLANTIC

There is no need to flap or flutter. The great flier keeps his wings spread taut, the movement of the warm air occasionally creating little tremors along the great expanses of skin stretched between arm and body. There is nothing at all to cause concern on this clear, calm day. *Quetzalcoatlus northropi* is the largest airborne creature on the planet, too big to be hassled by the smaller squawking birds and pterosaurs, too high aloft to worry about the shadows of the hungry mosasaurs swimming under the waves below.

He's heading home. Not that he carries the concept of a permanent residence. But there's something in his reptilian brain that urges him to go back to where he came from, to return to the flat, fern-covered plain where he had hatched years before.

There will certainly be others there. *Quetzalcoatlus* come together to squabble, mate, and build their nests en masse, dozens of nests spaced about a wingspan apart through the lowland. The acrid guano stench is terrible during such gatherings, but that's unavoidable. Tyrannosaurs pick off any *Quetzalcoatlus* that nest alone or too far from the margins; their eggs and hatchlings are relished by the parrot-like dinosaur *Anzu*. Together, some of their number might be lost, but not even a full-grown *T. rex* knows how to cope with a wall of screeching, pecking pterosaurs shambling toward them. The great pterosaurs will survive together or not at all.

The *Quetzalcoatlus* held aloft over the sea is certainly a survivor. After a few close brushes with hungry *Acheroraptor* when he was little more than a hatchling, he managed to peck his way to the top and grow to an adult with a wingspan about thirty-three feet across. For more than a decade he has followed the seasons as they have shaded across the Northern and Southern Hemispheres, stopping here and there to pluck a few small dinosaurs out of the low-lying ferns before soaring off to the next buffet. If there is any creature that knows the Late Cretaceous world, from its swamps and forests to its mountains and seas, it is *Quetzalcoatlus*.

The soaring pterosaur clacks his jaws. His stomach is feeling empty. He will need fuel soon, but there is nothing to do about it now. If he dips down to the level of the water, he may very well be snatched by one of those giant seagoing lizards or a shark with rows of serrated, triangular teeth. He'd seen that happen once before to a smaller pterosaur who got too close, a beautiful flight suddenly interrupted by a splash of foam and a broadening circle of red. And even if he were able to find a patch of sea devoid of predators, his jaws are not evolved to pluck up fish on the fly. Water is far denser than air, after all, and dipping his lower jaw down into the waves after a flash of silvery fish scales would cause so much drag that he would fall head over feet into the water, where he'd have to struggle, flail, and flap to get back into the air, a vast waste of energy compared with whatever few oily calories he might gain.

Better to wait for the mainland. There might be some yearling dinosaurs to pluck up and swallow. He's become a bit of an expert at catching them. While certainly ungainly on land, *Quetzalcoatlus* is still able to stride as if on stilts—one of his steps matches many of small, fleeing dinosaurs. His long neck and beak help, too, closing the distance to a savory dinosnack that he bashes against the ground before swallowing whole, having no teeth to chew with. A few of those might give him enough energy to get all the way back to where he hatched. He'll be there soon. Not too much longer now.

3

The First Hour

The *Ankylosaurus* senses that something is very wrong.

Being a living tank has shielded *Ankylosaurus* from a great deal through the two decades that she's been shuffling around Hell Creek. The rows of keratin-covered scutes on her back, the pointed spikes jutting from the anatomical seam between belly and back along her sides, and, of course, the knob of heavy bone at the end of her tail have provided enough of a visual warning that only the most daring or desperate tyrannosaurs cross her path. Even her eyelids are armored, covered in toughened scales that can prevent nasty pokes or abrasions. And in those times when a tyrant has ambled close enough to be a threat, all *Ankylosaurus* has had to do was fold her legs beneath herself, press her body to the earth, and swing that great club from side to side. Any carnivore who doesn't get the message from twenty-five feet of grumpy ankylosaur defending her space might very well leave with a shattered leg and an eventual death by sepsis, hunger, or the jaws of another predator.

But the ankylosaur's unease now is not from any tyrannosaur or other form of sharp-toothed interloper. Her enemy is everywhere and nowhere all at once. The ground rumbles and shifts, as if the very soil beneath her hoofed feet might be yanked away at any second.

Just moments earlier, the great *Ankylosaurus* had lumbered over to the edge of a lake dotted with cracked stone along the margins—a firm foothold where she could dip her armored head down and sip contentedly without worrying about getting mired in the softer, rank mud filled with decaying debris. No sooner had she taken that first cooling draw, however, than her footing seemed to slip and slide beneath her. Everything trembled, as if all her dinosaurian neighbors had begun jumping up and down all at once. But she could see no other large dinosaurs nearby. Save for a few birds and circling pterosaurs overhead, she was alone. To her flank, where the forest jutted out along the lake edge, the trees seemed to dance back and forth, jerking and swaying, a spray of squawking birds shouting their irritation. One tall *Metasequoia* toppled sideways, spraying soil into the air as the dense dawn redwood felled a dead nearby tree with a great crash.

Then, soundlessly, the shaking stopped. The trees swayed back to their upright position as if nothing at all had happened. All was suddenly, eerily quiet. The *Ankylosaurus* stood still. She had no comparison for what just occurred, no experience to guide her in what to do next. She stepped back from the cloudy water's edge and held still, tail aloft. She waited. She breathed. Nothing. A pterosaur far above let out an annoyed screech at another that ventured too close, but there was no other sound. That itself was scary. Hell Creek could be a noisy place. The hoots, honks, growls, squawks, and other miscellaneous sounds rarely left a silence. Even in the dark of night, insects stridulated and called to each other through the undergrowth. Now everything had stopped. The entire Earth seemed to wait. The *Ankylosaurus* snorted and began to turn. She would go back to her favorite shady spot, a comfortable divot she had dug into the soil within a small copse of conifers. That's when the second tremor jolted and shifted the stone beneath the soil, the geological conductors of the seismic waves that were coming in pulses, just minutes apart. The defense that had served her so well, the strategy that had seen off the worst and most terrible of rotten-breathed carnivores since she was a yearling, seemed foolish now. What could she do against the earth beneath her feet?

No two impacts are exactly the same. The speed of the impactor, its size, the angle at which it burrows into the planet, and the type of environment it strikes all have roles to play. Which is to say, the sequence of events that is just now reaching Hell Creek was not inevitable. With just slight differences to the Earth's rotation or the asteroid's speed, the great chunk of stone might have burrowed into the ocean. In that case, the crater still would have been massive, but having to plow through so much water would have at least slightly buffered the asteroid's devastation. Or the bolide might have hit ground underlain by rocks created during the Carboniferous, the great coal age when vast swamps filled with early trees were buried and turned to immense carbon seams. In that case, the impact may have ignited the underground coal deposits and started subterranean fires that would have dumped vast amounts of carbon dioxide into the atmosphere. The resulting global warming from such a scenario would have been a difficult circumstance, certainly, but one that life on Earth had coped with and persisted through before. Then again, the asteroid could have hit relatively inert stone. In such a strike, rock would have been melted, the local area would have been shaken, debris would have been thrown into the sky, and perhaps the impact would have triggered nearby volcanic systems to suppurate lava over the surface, but the disaster would have been local.

As things stand, however, the situation can hardly be worse. Too fast to even see as more than a streak of red-hot light, the asteroid had approached its strike point from the northeast. This wasn't a perpendicular hit. The mountain of dense stone rammed home at an angle of about 45 degrees. One moment everything was as it had always been. Then, in a blinding second, this spot on the Earth's veneer burst open like a popped pimple. The force of the impact was so great that the numbers describing it veer off into the ridiculous. The impact released as much energy as 100 teratonnes of dynamite, 420 zettajoules worth of energy, a collision so incredible that there simply is not a way for any mind on the planet to truly understand the scale of this crash. Not since the early formation of Earth itself has there been an impact like this, the one impossible chance. Think of the largest explosion you can think of, and it is still absolutely dwarfed by this impact.

The offending asteroid is not like some small fleck of space stone that plonked down in the middle of the field somewhere. The bolide did not just bury itself into the dirt and helplessly steam. The driving force of the impact plunged the asteroid into the Earth's crust, the hardened skin of the planet immediately melting from the heat and pressure of the cosmic smack. The stone did not just break as the bolide made contact. Rock that was not immediately vaporized or ejected out into the atmosphere turned to an almost liquid form, stone moving like viscous blood as the rock was pushed away by the force of the punishing impact. Think of the way a droplet of falling water creates an indentation in the surface of a body of water, forming a tiny, sparkling ring that rises and then folds back over the place where that droplet hit. Now imagine that the droplet, and the water it plunked into, was rock, not in a puddle but over an area large enough to be seen from our moon. Rock was pushed away and up from the point of impact, the towering sides still too liquid to hold their own weight. The stone rose and fell back toward the center, creating a middle peak over the strike point. But all this pummeled, shattered stone remained in motion. No sooner had the middle peak of the crater formed than it began to collapse in on itself. The rock cracked and broke and fell, nothing more than a disturbed pile of jumbled stone.

All of this happened within five minutes of impact.

Destructive as it is, forming a crater more than a hundred miles in diameter, the moment of impact is only the flashy beginning. All that force has to go somewhere. And it ventures out, conducted through stone and water as waves of force inadvertently announce the impact around the planet.

Where the asteroid burrowed in was along the edge of what would one day become the southern part of the Gulf of Mexico. This is a coastal strike, affecting the sea as well as the land. Untold gallons of seawater have been immediately vaporized from the heat of the impact, creating a great empty space in the ocean. On the edge of this void, the water has been pushed away in immense mega tsunamis that spread outward, waves hundreds of feet high racing toward shores where even the friction and drag of the land—its beaches, its forests, and all its living things—stop the waves only after they reach miles inland. Bodies of great sea reptiles caught in the waves are slammed down onto what had just been dry land, water roaring and churning over environments where dinosaurs roamed. Then the water rebounds. Hitting the shores so forcefully reflected the waves in the direction from which they had come, racing back toward the site from which they had been pushed away, carrying debris and sand with them. By the time the waters calm, over three hundred feet of sand are piled on top of the impact site.

The force moves through stone, too. Solid rock is not nearly as immutable or immotile as it seems. It just keeps different time, slower time. A great enough shock has moved rock into action, shifting and straining as the physical energy of the impact is absorbed by the Earth's crust. The force of collision doesn't just drive down into the Earth's outer layer and slosh stone around the strike point. The force flows outward, moving through the bedrock across and through plates. Mile after mile, the force spreads through stone, heading all the way to

Hell Creek. The time it takes to travel there? Around fifteen minutes.

The first pulse had taken the *Ankylosaurus* by surprise. The second was a strange and inscrutable nuisance, a great shake-out that toppled trees and made dinosaurs trip midstride. The third pulse, arriving within ten minutes of the first, has graver consequences.

When the *Ankylosaurus* arrived by the lakeside, all had been calm. The lake was still, save for the movement of water striders and the occasional ripple of a fish trying to nab an insect snack from the surface. Now the water seems to have a tide, but one that is going much too fast. All the shaking began to slosh the water back and forth inside the basin, the water picking up more momentum with each swing. To one side and then the other, back and forth, water holding tight to water until it seems like the entire lake is moving in one great sweep. And then it spills.

The weight of the water has become too great for the basin to contain anymore. The wave rises to one side of the lake and then falls, scraping against the bottom with such force as to dig up ancient sediments that had been laid down long ago. Ten million years earlier, this place had been part of a warm inland sea. Coil-shelled ammonites zipped through the sunlit waters, trying to avoid the crushing jaws of mosasaurs. The cephalopods left as the waters receded and drained off the continent, the dead becoming fossils that lay beneath the ground for millions of years, unseen beneath the lake. But now these ancient and empty shells have been brought into the sunlight again, dredged up by the seiche wave within the broad basin. The water scrapes along the bottom, trips over itself, and crashes, sending another wave out over the surrounding landscape—right in the direction of the *Ankylosaurus*.

Suddenly six tons of ankylosaur are little more than an enormous top, spun and battered within the water. The dinosaur's great, osteoderm-studded back strikes the trunk of a tree while she is tossed by the turbid waves, sending her hurtling in another direction. She flails her limbs, trying to find anything to gain purchase. *Ankylosaurus* did not evolve to cope with earthquakes and violent inundation. She had evolved to defend herself from sharp teeth and munch ferns as slowly as she pleased. She is about as streamlined as a brick, and there is nothing she can do as she is tossed, part of a soupy mix of dinosaurs, dirt, plants, fossils, and debris spilling out as far as the water can carry her.

She awakens soon after the dousing. Somehow the waters have funneled her down a channel, bringing her armored heft to rest on a sandbar. She tries to stand, but a lick of internal fire shoots through her left hind limb. Something is broken. She tries again, letting out an agonized hoot as the pain flares again, but, even if slowly, she can at least propel herself along on three legs until she finds a place to rest, sleep, and begin to heal.

Everything feels strange now. Entire groves of trees have been pushed over by the floodwaters, the hold of their roots perhaps weakened by the earlier tremors. And the late afternoon sun shines red, a large angry orb on the distant horizon. The *Ankylosaurus* has seen this before. Days when the forest burns turn the sun redder than the flames. She has no understanding of why, the cause and effect of soot and dust changing the way the sun's rays come through the atmosphere, but knowing the connection is enough.

Despite being drenched, however, the ankylosaur doesn't feel cold. If anything, the late afternoon air feels preternaturally warm. The heat pervades the air itself, making distant points on the floodplain seem to shimmer. It's only going to get hotter.

What goes up, must come down. The asteroid that careened into the planet did not simply rest with a great *wumpf* as a whole chunk. Pieces of the stone came off as it entered the Earth's atmosphere, and even more rock was cracked and shattered as the bolide hit home. The stony strata of the Earth's crust, too, broke and pulverized—not just into boulders or rocks but to dust. The impact shocked the stone down to the mineral level, hardened quartz crystals battered from impact. The heat was so great that some of the stone turned to tiny glass spherules. The asteroid was the parent rock that inadvertently spawned uncountable numbers of tiny offspring, many of them thrown high into the atmosphere by the sheer force of the strike.

It's difficult to overstate the sheer amount of material this single event tossed into the atmosphere. The strike almost instantly created more than 12,000 cubic *miles* of pulverized, melted, distressed rock. The transformation of the underlying stone was so fast that the Earth-shaking hit threw more than 350 billion tons of sulfur and 460 billion tons of carbon dioxide into the atmosphere. The makeup of the upper atmosphere, now injected with hundreds of billions of tons of climate-altering gases, almost immediately begins to change.

All that dust, that glass, that debris from the bolide's collision do not simply settle like a blanket over the impact site. The products of the collision have been shot up high and now have begun to gradually fall back down to Earth. By itself, any one particle won't even register as a nuisance. But much like the stone that had created them, they create friction in the air as Earth's gravity pulls them back down. The amount of debris created by the impact is immense, enough to send products of the collision zipping off as far as Cretaceous New Zealand. And as all these little pieces of dust, rock, and other geological bric-a-brac do so, they all begin to heat up. They all rub against the air, generating heat as they fall.

Nearly an hour after the terrible tremors, tiny specks begin to fall in Hell Creek. Most are so small as to be unnoticeable. Some are large enough to cause little *plink-plonk* noises as they strike the ground, dinosaurs, rivers, trees. Their speed is not as great as their parent's. No dinosaur is going to be shot through with tiny holes. The terror is slower, harder to escape. The local temperature is beginning to ratchet upward, from hot, to the hottest it has ever been, to unbearable. Animals that can burrow are already doing so. Creatures of the lakes are submerging and sticking as close to the bottom waters as they can. Birds and small dinosaurs are seeking out what shelter they can within tree holes and abandoned underground dens. But soon even the forests won't be able to provide any refuge. The stinging flecks begin to catch the dry leaf litter of the forest understory. Innumerable blazes begin to burn and singe. Tongues of fire are starting to join each other. Walls of flame are starting to build, turning the remnants of the day into a premature, overheated night.

Ankylosaurus is panicking. Not far from where the flood deposited her, she nonetheless eases herself into the waters of an eddy, her hoofed feet digging into the silty mud to help hold her in place and keep her nostrils above water. She can't submerge. She can't sleep through this. A tyrannosaur, its fuzzy filaments blazing, screams and stomps past, disappearing in a direction where there is only more flame.

Elsewhere
OFF THE COAST OF ANTARCTICA

Air. *Morturneria* has spent his entire life beneath the waves, never knowing the touch of coarse beach sand on his flippers, but he will always be connected to the surface. Every short while, he has to raise his great elongated neck up through the water column to snort out salt-laden mucus and take another lungful of air.

Morturneria is not one of the largest plesiosaurs, nor even the most fearsome. For over a hundred million years, reptiles like him have proliferated into hundreds of variations through the world's seas. Some have been monstrous—bulky, short-headed predators that gorged on whatever creatures they could catch in their impressive mouths. But many shared his body shape, something along the lines of a snake threaded through the body of a turtle. His relatively compact body is the anchor point for four broad flippers, the wings that propel him through the water, as well as his ridiculously long neck.

Most long-necked plesiosaurs are piscivores, for the most part. They hunt by ambush, moving their bodies one way and their necks another to nab unsuspecting fish and squid. But this takes a great deal of energy. The open ocean is a world of feast and famine. Perhaps a plesiosaur would happen upon a shining school of fish or a flashing squadron of squid, or perhaps all they'd find would be the snapping jaws of a predatory mosasaur. But there are other food sources much closer to shore. Shells, crustaceans, and other morsels are splayed over the sandy bottom of the coastal shelf, a perpetual invertebrate buffet. The most abundant source of sustenance is not to be found in the treacherous waters of the open sea, but snug in the sand.

For millions of years before the time of *Morturneria*, plesiosaurs plowed through the sand along the coasts. Swallowing rocks, along with invertebrates, helped pulverize the shells and release the nutrition inside the shelly tidbits. But the earliest plesiosaurs to do so missed out on a great deal of food. Their snaggleteeth interlocked, but there were still broad gaps between each enamel-covered cone. Their oral nets could scoop up a bivalve or a crab, but the smaller animals easily escaped in clouds of silt the reptiles left behind. Those reptiles with smaller and more densely packed teeth were able to nab some of the small amphipods, shrimp, worms, and other goodies from the bottom sands; as a result, they had just that much more energy to find mates, reproduce, and pass on their peculiar smiles to the next generation. Repeated over and over again, this series of evolutionary happenstances led to *Morturneria*.

Lungs filled, the plesiosaur flaps his wide pectoral fins and makes for the bottom. He isn't in a hurry. If he were, he could move both his fore and aft flippers to speed away from danger. As it is, there are no sneaking sharks or shadows of mosasaurs nearby. He lets his hind flippers act as fleshy rudders as he angles down, searching for the best spot to dredge. *There.*

The teeth of *Morturneria* are not nearly as thick or as fearsome-looking as those of his plesiosaur relatives. Natural selection has led his lineage to have small, closely packed teeth—a toothy version of baleen. Nearing the bottom, he opens his jaws and flaps his flippers to propel his body forward against the drag created by contacting the sediment, scooping out a mouthful of sand, silt, and hidden invertebrates. Quickly, his jaws snap shut, his flat reptilian tongue pushing up against the roof of his mouth to help press out the fine-grained sediment as the little organisms become trapped in the net of needlelike cones. With a gulp, the mix of little meals start their long journey down the great neck toward the stomach. *Morturneria* turns to run the same underwater flight path a second time. Then a third. Then a fourth. For a few moments, this patch of seafloor is zigzagged with tiny ruts—runnels that will soon be covered over by shifting sands and play host to future breakfasts, lunches, and dinners.

Sated, at least for the moment, *Morturneria* returns to the surface for another breath. But not just that. From time to time he likes to spy-hop and raise his neck above the surface of the water. Sometimes it's just a matter of curiosity. Other times, it's saved his life—enough to detect the spout of an approaching mosasaur that may have otherwise snuck up on him. Water falls off in great sheets from his mottled blue skin as muscles at the base of his neck contract. The sun is getting low now, taking on a reddish glow. And then, something *plink*s into the water nearby. Resting at the surface for a moment, *Morturneria* turns his head to see another small ripple on the ocean's surface, then many more.

A small, hot pain stings the plesiosaur's head. He hisses, not sure where the source of the pain has come from. Now he is beginning to feel the same stings on his back. These are not the teeth of a predator or rival, and they don't hurt enough to send him immediately swimming away. He hisses again, something small and hard ricocheting off one of his teeth. *Morturneria* shakes his head. Better to dive below until this strange shower ends. He takes a deep breath and dives under the surface. The stings cannot reach him here.

4

The First Day

There is no dawn on the first day of the Paleocene.

From the ground, looking upward, the sky is black. But this is an illusion. The early morning sky—dotted with stars, the glowing moon reflecting the sun's rays—is entirely obscured. A black shroud has been drawn over Hell Creek, billowing clouds of soot illuminated from below by a conflagration that even fire-tolerant plants cannot withstand. The roar heard all across the Hell Creek is not the defiant shout of a carnivore or *Triceratops* vying for dominance. It is the unearthly roar of fire and the consumption of everything that it touches. The rising sun is completely blocked by the choking smoke that continues to billow as acre after acre goes up in frantic gouts of flame.

Little *Mesodma* tries to sleep through it all. A little ball of fluff, squirrel-like in both size and appearance, the tiny beast is curled up tight, head resting on the soft, rust-red fur of her tail. In technical terms, she's a multituberculate. She's classified as such because of her peculiar dentition—her cheek teeth have many ridges, or tubercles, that are far more ornate than those of many other mammals. The paleontologists who will come to study them will call them multis for short, little beasts that first appeared in the Late Jurassic—over 80 million years earlier than our present moment, during the time of *Allosaurus* and *Stegosaurus*—and flourished alongside the terrible lizards. In her case, her lower premolars look like broad, flat spoons covered in long ridges. Such teeth are excellent cutlery for biting through hardened foods like seeds, nuts, eggs, and even bone, all of which were in abundance just hours before. Now, she's not only among the few mammals left in this part of the world—she's among the few animals left alive in the conflagration at Hell Creek.

Dark lumps of all shapes and sizes dot the singed clearings and blackened, smoking forests. Some of them are charred, the remains of large dinosaurs hemmed in by the fire. Darkened, scaly skin peels away from the frill of a downed *Triceratops*, not far from a similar lump that was once an *Anzu*—a beaked, feathered dinosaur that now looks like a strange roast turkey. Another prone form is that of *Didelphodon*, one of the largest mammals of its time. About the size of a house cat, this marsupial was a consummate egg thief and hunter of smaller lizards. But in all that scuttling about through the Cretaceous world, the little beast didn't have a refuge from the fire. Out on the surface, nothing stirs except the ongoing wildfires.

The fires are only the insults to a deeper ecological injury. Out in a marshy clearing, the sinuous form of an iguana-like lizard rests not far from the body of an adolescent *Tyrannosaurus*. There are no signs of fire on the dinosaur's body, not a filament or scale blackened. But the terrible heat had been too much to cope with. The re-entry of the impact debris created a global infrared heat pulse that lasted for hours. That energy alone was enough to raise the air temperature to 500 degrees Fahrenheit. The air itself can burn.

Finding shelter from such searing heat is almost impossible. It isn't just that the air has turned hot and stifling. It's that the impact debris has burned so hot, so persistently, that each tiny speck is contributing to a global infrared pulse that is lighting up the entire planet. In places not blanketed under choking smoke, there are no shadows. The skies are so bright now, burning with impact aftermath, that the light has turned oppressive.

Even on a normal day, prolonged and unsheltered exposure to the sun can be uncomfortable—it can even burn. Now that intensity is ramped up to ten times the power of a normal day's sunlight, heat that literally blisters and singes whatever organic matter it touches. Sheltering in a valley or crevasse won't make a difference, nor hiding under the shade of a tree. Even in the waterways of Hell Creek, the oily, water-resistant feathers of waterbirds have begun to curl and singe from the awful temperature spike, stripping away the hydrophobic properties of the plumage and causing these birds to take on water. The birds that have tried to stay in the water—able to hold their breath for only a minute or two—have either died from the heat or drowned, unable to struggle against the weight of the water now soaking into their feathers. The only waterlogged birds that stand a chance are those that have quickly waddled away to rock overhangs or other opaque shelters, the mildest of reprieves from the light and heat.

For many of Hell Creek's inhabitants, there is not even the slimmest possibility of escape. Non-avian dinosaurs, in particular, had evolved to live out in the open. There is nothing in their behavioral or anatomical repertoire that can give them even the slightest chance of survival. What could a dinosaur do against such oppressive heat? *Tyrannosaurus* and *Torosaurus*, *Edmontosaurus* and *Ornithomimus*, they all perished. The armor of *Ankylosaurus* and *Denversaurus* did nothing. The domed heads of *Pachycephalosaurus* and *Sphaerotholus* were irrelevant. The great hadrosaur herds could offer no protection, nor the brooding behavior of nesting troodonts. Almost nothing in the entire set of dinosaurian behavior matters anymore. Tens of millions of years of evolution, undone in mere moments.

Mass extinctions play out with such devastation because they are rare. They are, literally, the worst-case scenarios. In the past, such catastrophes always unfolded slowly and relentlessly. Change requires that organisms move, adapt, or die, and mass extinctions are built on the widespread failure of the first two options.

Those organisms that survive such pressures owe their success to pure luck. With the exception of bacteria and other fast reproducers that thrive through variations created by generational mutations, organisms on Earth can't evolve fast enough to adjust to dire situations in even the most protracted disasters. The survivors are those that, by sheer accident, possess traits or behaviors that allow them to survive. These creatures are rarely the largest, the flashiest, or the most charismatic. Many are generalists or the common species found almost everywhere. In a sense, the meek inherit the Earth—and they had done so four times before.

Now all that stress, all that destruction, has been compressed down into a single day. Life has never had to cope with a day like this before. What creature could possibly evolve to handle blast-furnace temperatures that have been roiling for hours? This is long enough to be deadly, but not long enough to adapt to, or to fortuitously produce a new generation with just that much more heat tolerance. Even then, what creature can withstand air temperatures approximately five times higher than even the hottest possible day?

Dumping heat had already been a problem for non-avian dinosaurs long before impact. Even though dinosaurs thrived in the endless Mesozoic summer, living large didn't necessarily come easy. The early success of the dinosaurs—part of their evolutionary secret that allowed them to so thoroughly dominate the planet—now had become part of their liability.

Physiology differs from species to species, but, on the whole, dinosaurs are endotherms. They generate their own heat supply inside their bodies. That's a big advantage when you're small, as the first dinosaurs were. They did not have to bask in the sun, stride around hot habitats, or otherwise get their warmth by soaking it up from environmental sources. Having a hot-running metabolism was a key part of what allowed dinosaurs to compete with ancient crocodile cousins and step over the snuffling little mammals. But in an animal as large as *Tyrannosaurus*, *Triceratops*, or *Ankylosaurus*, heat can be something of a problem. Bigger bodies tend to trap heat and have difficulty dissipating it. Some dinosaurs—those of the saurischian family, like *Tyrannosaurus, Alamosaurus,* and birds—have air sacs in their bodies that act almost like swamp coolers. The airflow through these pockets allows these dinosaurs to dump excess heat, even on sweltering days. Other dinosaurs wallow in mud and water holes, or remain relatively still through the hottest parts of the day. Dinosaurs had evolved to match their habitat, and until now, Earth had changed slowly enough to let them adapt as needed. But this was too much for anything other than a bacterial extremophile to withstand. Caught out in the open—with no underground refuge or other shelter—most non-avian dinosaurs died within hours of the asteroid strike, no matter how near or far they were from the impact point.

Belowground, *Mesodma* twitches her little paws in her sleep. The scene is absolutely serene compared with the firestorm above. The world she'll wake up to will be fundamentally different from the world as it was the day before. But at least she'll wake up. She'll have the soil to thank.

Shocking as the immediate aftereffects of the impact have been, the devastation is not omnipotent. There are places the effects don't reach, or at least are muted. While the asteroid itself buried deep into the Earth's bedrock and sent tremors racing through stone over thousands of miles, even a thin rind of soil is enough, just enough, to protect creatures that burrow beneath the ground. Even at their most intense, burning more than 800 degrees Fahrenheit, the heat of the first Paleocene fires reach only a few inches into the soil. The ground here was wet prior to the impact, part of the great coastal plain left over from when this area was covered by sea. The trees and leaf litter were exposed enough to dry and could do little against the falling debris and subsequent heat pulse, but the soil was effectively a moist blanket beneath the feet of the dinosaurs. The damp earth was the interface that the great dinosaurs could not access, but it was able to provide refuge to those that could.

And *Mesodma* is an accomplished digger. Her ancestors had been plowing through soil for millions of years. That's how they were able to avoid the ravages of dinosaurs while living right under their feet. In better days, her burrow was on the edge of an *Edmontosaurus* nesting area. She'd been born into this burrow, one of many pups reared by multiple families of the reddish-white, spot-dappled mammals. For most of the year, they foraged from the surrounding forests, plucking fruits when they could get them and gnawing on marrow-filled dinosaur bones. But the nesting season was when the real feast began. It could be a dangerous time. An unwary *Mesodma* might get squished beneath the foot of an unaware dinosaur, and the glut of food attracted the small raptors that sometimes sniffed out burrows and used their large, terrible claws to dig deep and pluck out a furry snack. For the most part, however, the mammals slept during the day and emerged at night to crack into the gooey, delicious *Edmontosaurus* eggs. Even one was enough to last for a few days, and the feast went on for months. And when the baby dinosaurs began to hatch, they were so unwary that a crafty *Mesodma* might be able to nab one and try the dinosaurian delicacy back in the safety of their den. The dinosaurs never wised up to these depredations. Each nesting *Edmontosaurus* created a clutch of about a dozen softball-sized eggs. Multiply that number by twenty or more *Edmontosaurus* and there was plenty to go around. It wasn't so much an evolutionary stalemate as mammals taking advantage of a long-standing dinosaurian trait. Faster and more prolific reproduction allowed dinosaurs to outpace mammals, but that also meant that mammals could dine on dinosaur omelets as they pleased without the risk of driving their food source extinct.

Little *Mesodma* stirs at a snuffling sensation along her fur. It is another of her species, a young male pup that had been born to one of the other *Mesodma* in the burrow. Unlike most mammals up to this point, he didn't hatch out of an egg. For tens of millions of years, ancient beasts carried on the reproductive tradition of their reptilelike forebears and laid tiny eggs that hatched into pink, squirming little babies that lapped up milk that oozed from pinhole-sized openings on their mothers' bellies. Some mammals, in fact, still reproduced the same way. *Mesodma* females, however, kept their eggs inside. Offspring began to develop within their mothers, but they had to be born early: the small size of multis came with relatively narrow hips, too tight a squeeze for offspring to gestate for very long. Instead, the tiny, clutching babies were born early and climbed into the safety of their mother's fur. Warm, safe mom doesn't have a fully enclosed pouch to clamber into, but she does have a fold along her belly to keep her infants snug. She hardly eats during this time; the calcium within her bones starts to dissolve into her bloodstream in order to feed the growing infants. It's a legacy of why bones evolved in the first place, back when all vertebrate life was restricted to the sea. Ancient fish needed minerals from their bone-like armor to power the muscle contractions needed for long journeys. That same ability was co-opted later by mammals, bones slowly self-destructing so that underdeveloped infants with bones mostly made of cartilage can drink up the calcium they need to grow their own bone tissue.

In this case, the male does not need to nurse. He is well past weaning, but sharing touch and body heat is a common mammalian pastime. Just like the dinosaurs, mammals run warm and are largely endothermic. All the same, mammals have been kept at small sizes for over 150 million years. They still heat up and cool down very quickly. The origin of cuddling, however, gave some beasts an evolutionary edge. Living in tree holes and burrows, out of sight of hungry reptiles, brought mammals into close physical contact. That shared warmth was a boon to off-spring that often have difficulty staying warm on their own. Staying safe led to staying warm, locked in through natural selection as a behavioral trait that has made mammals more mindful parents than some of the terrible reptiles roaming the world. Affection and survival have become intertwined. The female *Mesodma* shifts and curls up again, the young male coming to rest his rump against hers, facing away. These are supposed to be the daylight hours up above. It will be many more hours until the slumbering mammals poke their noses above the entrance of their burrow to sniff the air and check for the presence of any nearby reptiles. By then, the non-avian dinosaurs will be all but extinct.

Other saurians, however, are pulling almost the same trick as *Mesodma*. They're just doing it underwater. *Compsemys* shifts and settles a little farther into the debris and rotten vegetation of the pond bottom. It's safe here. There's something strange going on above. The air is far too hot, and it even started getting that way before the trees burst into flames like twigs in roaring fires. The turtle, like all the other reptiles in the little pond, dipped below the water when temperatures began cresting 100 degrees, only sticking their nostrils above the surface when their bodies demanded another gasp of oxygen.

Compsemys is a small turtle compared with some of its neighbors, only about a foot long. The Hell Creek rivers, ponds, and streams are rife with shelled reptiles of one type or another. Some are encased in hardened shells of plate armor, a home made of bone much like those carried by some of the earliest turtles of the Triassic. The others are soft-shell turtles. Great, sprawling sun-baskers, these are turtles with narrow noses and a different armor arrangement that give them a softer look. All rely on the water. The waterways of Hell Creek provide food and shelter in such abundance that the ravages of crocodiles and champsosaurs—at a population scale, anyway—are negligible.

Mercifully, the time of truly giant crocodylians has come to a temporary halt. The forty-foot-long *Deinosuchus*, capable of busting turtle shells open with a single bite from its huge, rounded cheek teeth, had gone extinct 5 million years before the time of *Compsemys*. And even though there are crocodiles with crushing jaws here, like *Brachychampsa*, they are much smaller, more on the order of ten feet long or less. The others, like *Thoracosaurus*, focus on fish rather than turtles, and champsosaurs do the same. These latter reptiles are crocodile mimics with long, fine-toothed snouts. They act much like the crocodylians do, but belong to an entirely different evolutionary branch—a case of convergent evolution, two different reptile families evolving to snag fish and other small morsels in much the same way.

Compsemys is a predator, too. The turtle's large, triangular head ends in a thick, hooked beak. The turtle does not have the speed of the fish-eating crocodylians. *Compsemys* focuses on precision and biting so hard that a swimming fish might suddenly be cleft in two. Which half *Compsemys* eats first is a matter of heads or tails. In fact, the turtle can even turn the tables on the crocodylians now and then. Big crocodylians start off as tiny hatchlings, after all, and in the spring any pond with a healthy population of *Compsemys* soon starts to see some of those little crocodylians disappear from the surface of the water.

Food is not the prime imperative for the turtle right now, however. It's air. While wonderfully adapted for spending almost every second of life beneath the pond's surface, *Compsemys* still breathes with lungs. The reptile will always be tethered to the surface. In calmer times, he can swim to the surface or even bask out on a log for a while, the heat of the sun helping old skin peel and slough off the turtle's platelike shell scutes. Right now, however, the air is full of smoke and soot and smells awful. Even with the cooling protection of the water, taking a breath from the surface is not pleasant. *Compsemys* tries to avoid the inevitable deep breath for as long as possible.

At least the turtle won't have to return to the surface for a while yet. Each single breath can last *Compsemys* forty minutes submerged, and even a little over fifty if he is very still. That's an impressive feat, but still not quite long enough when the atmosphere burns with a hellish heat. Thankfully, through the millions and millions of years the ancestors of *Compsemys* lived and thrived in the water, turtles have evolved another skill. In a pinch, they can breathe through their posteriors.

At the back of *Compsemys,* at the base of the turtle's tail, there is a small slit. This is the outer opening of the cloaca. It's standard equipment on reptiles. Crocodiles, snakes, lizards, champsosaurs, and dinosaurs—including birds—have them, too. The cloaca, meaning "sewer," is the single opening for the urinary, excretory, and reproductive tracts, a central spot where all those tubes meet. But the cloaca of *Compsemys* has something special. Around its edge are tiny specialized openings called cloacal bursae. They're absolutely filled with blood vessels, and that's what makes them so important. When water enters the little openings, oxygen diffuses between water and flesh. The essential molecule is taken up in the capillaries, supplying *Compsemys* with a little extra oxygen to extend his time on the pond bottom.

The technique isn't the same as gills. *Compsemys* can't stay down indefinitely. But on a day like this, every little advantage can make the difference between life and death. An organism doesn't have to be impervious. All luck can do is highlight the traits that organisms already have and give each creature just that much more time. Even though *Compsemys* can get only about 20 percent of the oxygen he needs from this alternate route, it's enough. He can go longer than an hour underwater without straining himself, having to stick his nose up only a few times during the hours-long infrared pulse. The fires still rage, and will for some time yet, but sooner or later that fuel will run out. All the turtle needs to do is stay below, snag the occasional fish, and make each breath last as long as physiology will allow.

The battle for life on the first day of the Paleocene is won and lost by little more than biological threads. Only those organisms that are able to find shelter—below the ground, beneath the water—have any chance. All others, from the largest *Edmontosaurus* to the smallest insect, perish. There is no behavior that can save them. Evolution prepared them for the world of tomorrow, and perhaps the day after, but not for this.

But the survivors have no reason to celebrate. The impact debris and infrared pulse are the great match strike of the extinction, but the fire will have to burn on its own time. And Hell Creek has never seen fires like this. Fire, just like precipitation, does not come in only one form. Fire burns and behaves according to the conditions around it, whether that's the low, hot, nearly invisible burn over hot embers or the great, orange flames climbing up to the tops of trees. In a clearing that once saw shuffling ankylosaurs, stalking tyrannosaurs, skittering mammals, and grazing hadrosaurs, a cyclone of fire now rips across the debris-strewn ground.

The initial Hell Creek fires set the conditions for firestorms. As the flames spread, grew, and became hotter, the voracious consumption of oxygen in these pockets that had once been forest created a pull of more air toward the fire. Breezes created by the fires themselves brought in more air, more oxygen, as the updraft from all that heat spiraled on and on, twisting pillars of smoke high into the air. Each swirling tower is a point of intensity, not only combusting and melting what the cyclone of flame touches, but drying out the surrounding environment to make it more likely those tissues and materials will catch fire. Bodies of dinosaurs, dead and drying out from the hours of already incredible heat, catch flame and burn, contorting the carcasses. As the tendons along the backs of the dinosaurs dry, they pull. The heads arch back and the tails go up, flesh burning into blackened and dried remnants of what the magnificent animals once were. This has been the standard dinosaur death pose for millions of years, a reflection of the animals' unique anatomy. These charred Cretaceous bodies will be the last of their kind to take the posture, all that might be left to testify of the day the world burned.

Elsewhere
THE ISLAND OF INDIA

The bones of *Jainosaurus* form a great circle among the charred vestiges of the forest that once fed the dinosaur's rapidly grow-ing bulk.

Still a youngster, the body of this *Jainosaurus* is about the same size as a cow. The dinosaur had survived its critical first year, and was well into its second, adding inches to neck and tail and ever more pounds as the sauropod browsed among the trees and ferns. When she got older, her diet started to change. The bigger she got, the more forgiving her metabolism became. She could get by on bulk, swallowing down great quantities of less nutritious food. But during her young growth spurt, she had to select the most nutritious plants and fruits. Each mouthful mattered, especially as she was stuck in a race she had little direct control over. Great carnivores stalked these same forests, and most small sauropods never made it to the first anniversary of their hatching day. Their prime defense was becoming too big and too troublesome to take down. All she could do was eat, and eat, and eat, living through each moment in the hopes that some looming theropod had not blended in with the forest shadows.

But now that future has been cut short. Nor does she have any close relatives to carry on her legacy. This is true in the evolutionary sense as well as the immediately familial. All over the world, the sauropod dinosaurs are no more. Their family has been utterly and almost instantaneously extinguished.

With the exception of hatchlings, there was no such thing as a small sauropod. Even on islands, where more limited resources caused some species to become dwarfed, they were still far larger than most other dinosaurs and certainly all mammals. From the days of the Late Triassic onward, these dinosaurs evolved to be giants. There was no evolutionary route for them to gain wings or make dens beneath the soil or even swim all that well. Their size was at least partly reliant on the complex system of air sacs that allowed them to breathe more efficiently and keep their bodies cool. Those air sacs also acted like internal air bladders in the water, making these dinosaurs poorer swimmers than the predators that sometimes pursued them. The dinosaurs had not been granted the proper traits to survive the crisis. Evolution carried these dinosaurs along through thick and thin for millions of years, but sooner or later something cuts down biodiversity without care or malice.

Almost all the other dinosaurs are gone here, too. Most life, in fact. No group emerged unscathed. All suffered losses, even in a place as far away as the island subcontinent of India. At this time, 66 million years ago, India has broken away from Madagascar but has not yet completed its collision course with Asia. There are no Himalayas yet. India is, or just had been, another dinosaur-filled spot of geography, moving along by the conveyor-belt action of continental drift.

But the asteroid impact was so great that there was no place on the surface of the Earth that was truly safe. Life on Earth had become reliant on oxygen in the atmosphere billions of years before. So when the air itself became deadly, there was little that could be done for terrestrial life. From Paleocene Argentina to Japan, from pole to pole, any dinosaur out in the open quickly died, and fires flared over most of the world. The impact did not just send burning debris a few miles away or to the nearest continent. Sparks for the global flame spread all over the planet. The ancient Americas saw the worst of it, partly because of their proximity to the point of collision, but the effects were so dire that even on the other side of the planet, extinction still arrived nearly as swiftly. And as the first day of the Paleocene closed, the terrible events of the that day had only set the backdrop for the creeping, seemingly interminable struggle to come.

5

The First Month

Hell Creek is a skeleton of what it once was. In many cases, this is literally true. The charred and cracked bones of dinosaurs are scattered where the terrible lizards fell across the transformed landscape. Many of them are pocked and incised with the tooth marks of mammals like little *Mesodma*. The little burrowers that managed to survive nipped and chewed at the tatters of flesh and exposed bones, sometimes making their shelters beneath the splayed rib cages of Cretaceous titans. An opossum-like mammal, an *Alphadon*, shuffles beneath the bones that once held up the frame of an *Edmontosaurus*, a small gaggle of fuzzy babies hanging on to her back. Each day, she forages to find the insects, lizards, and other tidbits that will allow her to keep producing milk for these Paleocene youngsters.

The forest is a ghostly vestige of what it once was. Some trees still stand, denuded of foliage. Dead, they keep the same poses they had in life, only now their brittle forms creak in the wind, each gust holding the potential to topple them to the ground. The forest floor is crisscrossed with dead trees, some angled over the tops of slowly decaying dinosaurs. A deep charcoal black is the primary color of the landscape now, a sharp contrast from the rich earth tones of the living forest. But what's left of the Hell Creek ecosystem is not monochrome. Low down, peeking out from small, moist patches of soil beneath the toppled logs, is a chorus of ferns.

The only verdure in the whole of Hell Creek only reaches a few inches above the soil, but each photosynthesizing filament is a reminder that life continues. The little fronds and curls of green create a verdant haze over the crisped landscape. It almost looks as if spring is going to return to Hell Creek, new growth already peeking out from the shreds of the Cretaceous. A little *Acheroraptor* makes a fern shake as it strides through the undergrowth, keeping an eye out for any telltale movement that might give away a snuffling mammal or scurrying lizard.

Even though more than half the Cretaceous species are now vanquished, not all the dinosaurs perished in the initial heat pulse and subsequent fires. Avians fared the best; some managed to burrow or find shelter from the infernos, emerging to find a world where the giants had all disappeared. And a few straggling non-avian dinosaurs survived, too. They were not the terrifying titans that cast long shadows over the landscape. The likes of *Triceratops* and *Tyrannosaurus* had been snuffed out forever. Earth had never seen their like before, and never would again. There was no escape for animals so big. But some of the maniraptorans, like *Acheroraptor*, were small enough to take advantage of underground shelters scattered around the landscape. Burrows dug out by turtles, crocodylians, and other creatures offered brief refuge to these small, feathery dinosaurs. One plucky *Acheroraptor* even got a lucky break, taking shelter in a little mammal burrow that was still being inhabited by its residents—a mammalian meal ready to eat. But such fortuitous moments that have survived are always fleeting. Even for those dinosaurs that survived, there was little to thrive on. And now the skies are darkening further.

Hell Creek wasn't the only place to burn. All over the world, forests became enormous tinderboxes as the post-impact heat intensified. Few reaches of the terrestrial world were untouched by fire. And each of those massive fires, in turn, sent plumes of smoke, soot, and ash high into the sky. Rising over fourteen miles into the air, the ghastly products of the fires were easily buffeted by wind currents to travel far and wide. And when the entire world is burning, the smoke quickly becomes all-consuming. Even as the fires abate, the skies remain dark. In fact, they only seem to grow darker—not from the billowing smoke of active burns but by debris creating a vast dome over the atmosphere. Sunlight has become a luxury.

But there's more to the legacy of the fires. Wildfires often produce very, very small particulates. They're so tiny that they are readily inhaled and can pass into lung tissues, making it harder and harder to breathe. There's no such thing as a truly fresh breath of air in these earliest Paleocene days. The surviving dinosaurs have a slight advantage. Their one-way respiratory systems, adapted to getting substantial oxygen from the air with little effort, allow them to get by. Other animals that rely on the great inhales and exhales of each breath find the air choking. Life isn't easy for any creature that dwells on or above the surface. Strangely, though, the fires contributed to another phenomenon that will act as a reprieve for much of the Paleocene's surviving life.

Fires release a great deal of carbon dioxide into the atmosphere. Billions upon billions of tons of carbon dioxide have been buffeted and billowed into the air by the global fires. And throughout Earth's history, large amounts of atmospheric CO_2 have fueled global warming—sometimes with catastrophic results. In this case, however, the opposite is true. The glut of the gas created by fires, and belched out by volcanoes in prehistoric India, will prevent the mass extinction from being even worse.

The surviving organisms on Earth can't know it, but the planet is just in the early phases of a long, terrible impact winter. Fire and heat have already had their day. There is little left to burn even if another heat pulse began in earnest. But now the world is about to shiver. Even weeks after impact, the compounding consequences of the catastrophe are just beginning to make themselves known.

Rocks are a critical player in the unfolding extinction. That may seem strange, as remaining organisms clutch hard to what's left, but life on Earth has only ever existed because geology allows it. The fate of stone is the fate of the world.

If its existence can be considered with any sense of vitality, or even a beginning and an end, the life of a stone unfolds too slowly for even the longest-lived organisms to comprehend. Consider a stratum of sandstone, tan sheets of weathered rock dotted with lichen. That stone was formed out of individual sand particles that were laid down on top of one another, compressed, and lithified. Sunk below the surface, raised into the air again by long-grinding tectonic forces, worn back down to become the sediment and soil of another era, the components and geochemical profile of the past providing the foundation for the present. Entire forests have grown from the wreck of ancient seafloors. Deserts expose what were once vast lakes. Each sheet of rock and broken boulder contains stories within stories, as unique as fingerprints.

The Earth is not covered in a single sedimentary rind as if it were some sort of cosmic orange. Igneous, metamorphic, and sedimentary, basalts and mudstones and gneiss, exposed in deserts and shrouded in forests—the rocks of the world are incredibly variable in their character, from their shape to their elemental components. Geology carries consequences.

The massive impactor that struck the planet at the beltline is part of the geological story. The stone was once part of an astronomic story, haphazardly catapulted across the solar system to meet alien rocks on our planet. And while we could certainly ponder what would have happened if the asteroid missed—sailing clear as so many have and continue to do— that scenario is too simplistic, too binary, as if there were only two possibilities on the fateful day—hit or miss. The stony truth is much more tangled. The unfolding devastation of the asteroid was not, you might say, set in stone. There were alternate possibilities for the fate of the planet if the rock had struck in the seas or driven into a different sort of bedrock.

There might have been local disasters, and perhaps even a global heat pulse, but the surviving species might have been spared what they are about to suffer. In such an alternate history, carnivorous little raptors might have continued to prey on the tiny squeaking mammals, using their scythe-shaped claws to burrow down and assure that—in some fashion—the reign of the dinosaurs would continue for millions of years more.

Under altered circumstances, a new Age of Dinosaurs might have risen from the surviving raptors. After all, back in the Triassic period, some of the first dinosaurs were bipedal carnivores that snatched up weasel-like protomammals and scraps of crocodile kills. Carnivory can often lead to omnivory, which is the gateway to herbivory. This happened over and over again during the Mesozoic era. During the Cretaceous, for example, a group of midsized predatory dinosaurs began to dine on more varied fare. Perhaps they couldn't compete with the other hunters of their time, or they just had more cosmopolitan tastes. In any event, these dinosaurs underwent a series of changes that played out multiple times among different branches of the dinosaur family tree. Natural selection began to favor dinosaurs with smaller, closely packed teeth. Dinosaurs that could digest the hard carapaces of bugs and extract the most nutrition from plants fared better. But eating plants requires that food move slowly through the gut to make the most of each bite, so dinosaurs that had guts capable of fermenting leafy greens prospered. Such guts required more space, of course—wider rib cages and hips. And those nipping teeth became even smaller, pulling back from the front of the mouth and transforming into a beak. These creatures were the therizinosaurs, great tubby herbivores that evolved from

carnivorous ancestors. They did not turn into herbivores overnight, but stand as proof that, given enough time and impetus,
the accidental nature of evolution can transform a carnivore
into an herbivore or vice versa. If the raptor-like dinosaurs had
been given a reprieve from the impact winter, they could have
set the stage for a second Age of Dinosaurs just as diverse, vibrant, fluffy, and scaly as the 135 million years that had come
before.

Such a second reign of the dinosaurs is not to be. The target rocks that the bolide struck had been formed in ancient
seas millions of years before. Known as marine carbonates, the
rocks at the point of impact were what was left of a warm,
shallow sea that hosted great reefs entirely packed with life.
Colonies of corals grew into tiny empires in the sunlit waters.
Small, armored amoebas known as foraminiferans proliferated
as plankton, along with disc-covered algae called coccoliths.
The rock that was eventually left behind by these waters was
not principally made up of sandy sediment, but was largely
composed of the shells these planktonic creatures so assidu-
ously grew to give themselves protection and shelter. Changing sea levels eventually caused the little patch of ocean to dry
up. The waters receded, the seawater evaporating and leaving
behind vast amounts of salt crystals—codified in stone as an-
hydrites. The principal chemical composition of these transformed salt crystals is calcium sulfate, $CaSO_4$. It's the latter
part that's most concerning for the creatures of the Paleocene.

Impact immediately transformed a great deal of what was
once stone into airborne particles. The strike not only released
vast amounts of greenhouse gases, but it also converted calcium
sulfate into an aerosol. Sulfates spread through the atmosphere

just as the impact debris and dust had. If we could zoom in on these tiny, tiny particles as they waft around the atmosphere now, they would appear translucent. They might even seem rather harmless. And in small amounts, they would be. But the impact created vast clouds of sulfate aerosols, thrown high enough into the atmosphere that air currents rapidly distributed them around the world. In such overwhelming amounts, sulfate aerosols reflect a great deal of incoming sunlight.

The sun's warming rays are not a given. They can reach Earth only because the components of the planet's atmosphere have traditionally allowed them to. But all the debris cast into the air and smudged all over the world have created a terrible, strangling scenario. Along with impact dust, the raging wildfires dumped astronomical amounts of soot and other dark, carbon-rich particles into the air. The dark color of these particles is very good at absorbing solar radiation. In another scenario—like intense volcanic activity—these carbon clouds might have helped warm the atmosphere in the years to come. In fact, such an event will happen in another 10 million years from our present moment in the Paleocene. Right now, though, these particles are dark enough to create vast shadows. Worse, the sulfates reflect the incoming solar radiation. Whatever light isn't immediately reflected back out toward space is absorbed, obscured by a global dust cloud.

Even as the fires ebbed, daylight has not returned. The skies remained dark—not as black as starless nights, but a hazy darkness that blocks out as much as 20 percent of the sun's incoming rays. That's more than enough to begin destabilizing the very underpinnings of almost all Earth's ecosystems—or at least all of them that are based on the constant production of photosynthesizers, be they trees or algae. In the evenings, as Earth turns away from the sun and brings light closer to the horizon, the sun takes on a menacing red shade—particles in the air scatter the blue and violet wavelengths of the spectrum, leaving red to dominate. This will remain so for some time to come. The sheer force of the impact propelled ejecta, carbon dioxide, and sulfates much higher than any volcanic eruption could ever throw them. Reaching the stratosphere, the particles have spread much more widely—and take much longer to fall back down. This isn't going to be a few months of darkness, or a year without a summer. The chill is only just beginning to settle in. The balmy days of the Cretaceous are over. In a matter of weeks, the temperature has plummeted. It'll keep falling. The average global temperature is set to fall more than 60 degrees Fahrenheit from its Cretaceous height, an inescapable chill regardless of season.

Against the looming darkness, the Cretaceous survivors do have an unlikely, unintentional savior. That might be a strange sentiment given all the horror that has already unfolded in Hell Creek and around the world. The death toll is already staggering, and the organisms that somehow survived the heat pulse will have to endure years of impact winter. Battered ecosystems will now have their foundations shaken as photosynthesis almost grinds to a halt, plants struggling to take hold. Perhaps for the first time in the history of life on Earth, though, volcanoes are coming to the rescue.

Far from Hell Creek, across an ocean and then even a bit farther, the Deccan Traps of prehistoric India are oozing. These are absolutely massive volcanoes, but they're not the sort that dramatically throw molten rock into the air like a geological sparkler. No, the Deccan Traps are flood basalts. They suppurate and flow, globbing lava over the landscape for hundreds of thousands of square miles around. And they've recently become active again.

More than a million years before the impact, the Deccan Traps pushed out a massive blanket of lava. Carbon dioxide and sulfates released from within the Earth came with the eruptions, causing a small drop in the global temperature. The dinosaurs of Hell Creek, in fact, may have gotten a little bigger than their predecessors to keep up with this change—the bigger the body, the better an animal is able to hold on to its body heat. The Deccan Traps then took a pause, quieting as the basalts they created cooled and took up new roles as part of Earth's surface. Until just before the impact. Just as the impactor happened to be getting close to its final target, the volcanoes started to pour more lava out on the surface, belching out massive quantities of carbon dioxide and other greenhouse gases in the process. This accidental effervescence is what will shield the survivors from the depths of the impact winter.

No single factor makes a decisive difference. There's a push and pull between the global geological and atmospheric disturbances that have been playing out since before the asteroid touched down. Soot, dust, the way sea ice reflects sunlight, sulfates, carbon dioxide, methane, and other players all buffer each other, and almost all are in motion, being thrown into the atmosphere, falling to Earth, or affecting the sunlight reaching the planet. And not all these factors are equally influential. Some have more punch than others. Carbon dioxide is an extremely powerful greenhouse gas, and the burbling output of the Deccan Traps raised the average global temperature by about 9 degrees Fahrenheit. It's a small concession in the broad picture, but it's enough. That's all that matters now. There's little opportunity to thrive. What life remains will have to get by on what's enough, waiting twenty years to get back to temperatures as they were before the asteroid struck.

The saving geologic grace of the volcanoes is out of character. In the past, volcanoes cut back life on Earth just as severely as the asteroid impact. The worst mass extinction of all, about 186 million years before the impact, was triggered by volcanic activity so dramatic that the output changed the relative amounts of carbon dioxide and oxygen in the air. The Deccan Traps eruptions have not been quite that intense, but, purely by chance, they are offering the world's surviving species a lifeline. And if the asteroid had never struck Earth, the non-avian dinosaurs would have been able to survive just fine. Dinosaurs originated in the wake of such eruptions, after all. The Triassic-Jurassic extinction, about 135 million years before impact, was triggered by intense volcanic activity in the middle of Pangaea. A warm pulse was followed by a cold snap, drastic temperature fluctuations that dinosaurs were able to survive because of their warm-bloodedness and fuzzy, feather-like coats. Maybe the Deccan Traps eruption would have led to the extinction of some species, but it wouldn't have eradicated the non-avian dinosaurs. It took the extremes of the after-impact world to decimate the dinosaurs, leaving birds and a few straggling raptors behind.

But no snow has touched Hell Creek yet. Whatever pale fluff is scattered about the habitat is ash, gray on the earth beneath gray skies. Patches of it are crisscrossed with small tracks. Some are five-toed—the marks of mammals that scuttle and scurry in this transformed landscape. Others are wide, splayed, three-toed marks left by foraging birds. But a few are little two-toed *v*'s, hallmarks of black-and-white *Acheroraptors*—like this one poking about for small prey, searching for something crunchy and wet among the slowly decaying debris.

Hunting is harder these days. Not only is prey less abundant, but the menu options have pared down from what they used to be and there's not as much cover as there was. There's hardly any point in hiding, stalking, and pouncing, descending on some little fuzzball with both feet, pinning the hapless mammal to the ground with a pair of sickle claws that sink deep into muscle and puncture viscera. These days all meals seem to come through a combination of desperation and luck.

But the slow burn of starvation hasn't quite set in yet. Just this morning *Acheroraptor* sniffed out a pungent, rotten smell from a den. It was a multi that had expired from one cause or another, beginning to stink up an abandoned burrow. It had to do. Meat is meat, whether fresh or going a little off, and this is certainly not a time to be choosy. *Acheroraptor* pulled at what was left of the mammal's musculature, his curved teeth sliding into the flesh and tearing backward as he took as many mouthfuls as he could find.

A rotting carcass can be quite a messy meal, though, and he could still smell some of the reek on himself. And while he can't give himself a tongue bath like the mammals can, he can at least preen a little. In a grove of dead trees, beneath one of the larger ferns, he sits, extends an arm covered in long brown feathers mottled with white, and begins to go through his grooming ritual. One of the feathers is broken and battered.

He bites at it, trying to hasten its death so that a new one can grow in its place. Most of the others are already clean, but then again, clean for a dinosaur often involves feather lice and other tiny parasites that have made the saurian their home. The literal hangers-on suffered a mass extinction, too. The loss of the dinosaurs has ended a great deal of the ecological activities that had evolved around them, holes in the landscape where these ecosystem engineers once lived.

The world of Hell Creek was absolutely dominated by dinosaurs. They seemed to fill the entire landscape, coming in every shape and size. They created a world that has no equal. The largest dinosaurian denizens of Hell Creek were over forty feet long and weighed nine tons, while the smallest were about the size of a chickadee, but the average dinosaur in the ecosystem was more than six thousand pounds. A subadult *Edmontosaurus*, all thirty feet of it, represented what the typical Hell Creek dinosaur was like. That's immense, and naturally those animals shaped the entire ecosystem around them.

Herds of *Triceratops* often followed the same paths to seasonal watering holes. Each time they marched along, they trampled the ground a little bit more. In time, over the course of years, they made entirely new wetlands that required more routes, which compressed even more ground even as some ponds filled with sediment and became land once again. When tyrannosaurs made nests, they scraped up vast piles of vegetation from the surrounding area. Not only did this amount to a bit of prehistoric weeding, but their piles of rotting, incubating vegetation also became homes for snakes, lizards, insects, spiders, and, especially after the nests were safely abandoned, small mammals. When ever-hungry *Edmontosaurus* and *Ankylosaurus* mowed down plants with their mouths, they shaped what would become of the forest. Young, juicy plants were always the best delicacy, so these dinosaurs often cropped off young plants before they could take hold. These megaherbivores kept the meadows and open ground clear, just as *Triceratops* did when they'd rub their horns against trees to the point of toppling some over. Soil was packed, seeds were scattered, carcasses were left behind to nourish the soil with all the chemical components of flesh and bone. And vast quantities of dung were left behind by everything from lone carnivores to massive honking herds. Dinosaurs did not merely inhabit the world as if it were a ready-made diorama. Dinosaurs literally made the world their own.

Of course life evolved around these dinosaurian forces. Insects easily accepted the terrible lizards as sources of both food and shelter. Fleas, some as large as an inch long, were a common source of discomfort in the forest. These insects evolved specialized sawlike mouthparts to quickly bite through scales and feathers to get their blood meals. Lice, of course, found plenty of room to make themselves at home among the complex scales and lavish feathers of the dinosaurs. And how could beetles get by without dung? All the Cretaceous plops offered a seemingly inexhaustible source of sustenance, not to mention a place to deposit the next brood. Even dead dinosaurs had their uses. Since the Jurassic, at least, some beetle species would gnaw their way through exposed dinosaur bone as if it were the cuticle of a leaf, eventually burrowing down through the spongelike tissue to create little crypts for their larvae. No part of a dinosaur went to waste.

But there were to be no more great piles of dino dung, and there are far fewer hosts for the parasitic species. There are no great warm blood-filled scaly tracts of topography to conquer. What lice remain are carried by the few survivors, avian and non-avian dinosaurs alike. They are what remains of their own mass extinction, not so much killed by the direct effects of impact but by the loss of their homes and sustenance. The lice that remain will have to jump between whatever feathery species they can latch onto and find a new rhythm in life's song, taking up on the birds that were able to hide from the worst of the heat.

The worms don't have nearly as much to worry about. Even if some near the surface were fried by the infrared pulse, most were sheltered by the soil. Their whole world is underground, a place less susceptible to swings in temperature and climate. And they reproduce fast enough, and in such numbers, that they are among the organisms that have relatively little to fret about through the coming winter. The worms are not gigantic, and perhaps they might even seem plain. They simply continue on, seeking out morsels within the soil, making horizontal burrows as they travel on. Just like the parasites and surviving insects, they have no understanding that anything is different, or that the world once was one way and now is different. They simply continue on.

Elsewhere

PALEOCENE ASIA

The forests are quieter now. Not just for the lack of non-avian dinosaurs, but the loss of the little creatures. The dawn chorus of birds isn't as raucous during each gray morning, and there aren't as many multis screeching and chittering over their territorial boundaries as there used to be. The mammalian screaming matches—one multi lashing its tail from the branch of a tree while another flags just as enthusiastically from a neighboring pole of pine—rarely happen now. The multis, except for a few, are almost all gone here.

But the mammals have not entirely disappeared. Low down on the charred forest floor, waddling along as if it had all the time in the world to do so, is a roly-poly little beast. This small creature is perfectly content to snooze in its underground burrow—evicting the occasional snake, of course—and wander out now and then to search for seeds, shoots, and nuts, taking a little of this and a little of that from the fresher parts of the Paleocene salad bar.

The mammal is a placental, a relative newcomer in mammalian terms. Even though mammals originated around the same time as dinosaurs did, placental mammals like this one have only been around for about 35 million years. Much like its other mammalian relatives, this trundling ball of fur gives birth to live young that are carried inside to term. The babies are still relatively helpless, eyes shut and requiring mother's milk for weeks, but they're up and about much faster than the offspring of the marsupials and monotremes, which also saw their way out of the disaster. But that's not the only advantage the little mammal carries.

Unassuming as it seems, this fuzzy creature represents the early beginnings of what's arguably the most successful mammalian lineage of all time—the rodents. This creature is, in effect, a protorodent, a small beast with a set of small, chisel-like incisors at the front of its mouth. These teeth will be the mammal's ticket to success.

The protorodent isn't the first mammal to have teeth quite like this. In fact, many multis have been gnawing away with similar front teeth for millions of years. The detailed dental equipment is what allowed those mammals to get a little extra calcium from chewing on dinosaur bones and generally make short work of even the hardest foods. But here in Paleocene Asia, the multis are almost all gone.

The Cretaceous world was not homogenous in its landscapes and species. Similar communities of organisms might be found on different continents, but made up of different species with varying behaviors and adaptations. While the multis of North America included some skilled burrowers, like *Mesodma,* those of Asia spent much more time aboveground and in the trees. They had not constructed the necessary refuges by the time the asteroid struck and the world burned. Most were lost, while other mammals—like this protorodent—survived almost unscathed.

In almost any other place on the planet, the multis might have outcompeted such a protorodent. Multis were already well adapted to doing the same things that rodents do. And their kind will survive here and there on Earth for some tens of millions of years more. But here in Paleocene Asia the multis all but lost. There is a great ecological gap where multis once were. The field is wide open.

The shuffling protorodent sniffs at the edge of a downed log. She urinates on it. Another one of her species has already been here, and it's in her nature to respond in kind. Perhaps an encounter would result in a fight. Perhaps it would result in the beginnings of another brood. Either way, keeping up scent signposts is just a mammalian thing to do.

She ambles on, eventually coming across the exposed seed of a tree. Perfect. In her tiny paws, she grasps the hard-shelled embryo and starts to put her teeth to work. It'll take time to get through the outside to the nutritious matter within, but she'll get there. Bite force isn't the only way to crack a nut.

Temperatures will drop here like everywhere else. The impact winter varies as much as the clouds do, but it is a heavy, shifting blanket over the whole planet. Every place is going to get colder. Through it all, though, these protorodents are going to keep eating, reproducing, and burrowing down into their warm nests, venturing out as they need to and hoping to avoid the claws of the surviving dinosaurs. For today, the name of the game is survival.

6

One Year After Impact

The cold keeps its grip.

The persistent chill contrasts with the fire-scarred land-scape around it. Hell Creek was not a steaming rain forest, but it was still a lush haven of greenery. Now all the green plants seem low to the ground, fiddleheads and hopeful shoots creating a forest that doesn't reach any more than a few feet high. The almost skeletal remains of forests still stand, tree trunks blackened by the fires that raged hours after impact, the slowly decaying bones of dinosaurs and other creatures still visible on the surface. Even as the soft parts have largely rotted away, save for hardened pieces of leathery skin that still stick to skulls and ribs, it'll be years before these mineralized parts of the terrible lizards fully disintegrate. The survivors of the Cretaceous world live in this vast graveyard, the remainders of ecosystems set against a reminder of swift death.

Dim, muted sunlight falls upon the bones of a *Thescelosaurus* that is slowly being pieced apart and broken down. For a "terrible lizard," this dinosaur might have been one of the most inoffensive saurians to have ever evolved. In life, this herbivore had been about ten feet long and weighed about five hundred pounds. That's on the small side for a non-avian dinosaur, especially in an environment where the hulking forms of *Triceratops* and *Edmontosaurus* were everyday sights. But life at small size suited *Thescelosaurus* just fine. The herbivore was an ornithopod, a distant cousin of the great duckbills, with a small beak and muscular legs to help the dinosaur outrun any predator that might come crashing through the forest. *Thescelosaurus* didn't have armor, horns, spikes, curved claws, or any other ornate dinosaurian finery. In some ways, it was a holdover from more ancient dinosaurian times, when some of the earliest herbivores filled the same niche. Perhaps *Thescelosaurus* was not the most dramatic part of the dinosaurian tale, but the reptile still was able to make a living munching on ferns and other low-growing plants; it was the Cretaceous equivalent of a deer.

But not anymore. Even if the dinosaur had managed to survive the initial fires and heat pulse, *Thescelosaurus* would have struggled to find enough food. It might still have wound up as a slightly disorganized pile of white bones, neck curved back over the spine and tail angled upward as more tiny cracks proliferated through the weathering bones with each passing day. The impact winter has fully set in.

While the weeks and months immediately following the impact and the great infernos saw dark skies clouded with soot, the persistent darkness has lifted now. Sunlight touches Hell Creek once again. But it's not the same as it was a year before. The sulfates that remain in the atmosphere, still stirred around by air currents miles up from the planet's surface, continue to reflect a great deal of the incoming sunlight back out into space. Together, they diminished the amount of sunlight reaching the planet by about 20 percent.

Looked at one way, the remaining denizens of Hell Creek can count themselves as lucky. Life on Earth depends on light, and most of the daily sunlight continues to reach them. The flora of Hell Creek have evolved to cope with annual seasonal changes as the orbit and angle of the Earth altered the length of each day. But a sudden one-fifth reduction in the amount of incoming sunlight is not something that they can deal with.

Plants are not left untouched by the calamity. Just as with the animals, the mass extinction pushed each species to the breaking point and beyond. Not only were there raging fires, but the intense cold of the impact winter has made it extremely difficult, if not impossible, for the seeds, nuts, and other surviving plant parts to grow. It's not a matter of space. The field is wide open to all comers now. It's a matter of sunlight.

The great and long-lasting success of plants relies on photosynthesis, a process billions of years old. It's the biological quirk that oxygenated the Earth's atmosphere, setting life on a pathway reliant on carbon dioxide and oxygen rather than on other gases. And just like those incredibly ancient photosynthesizing cells, plant cells turn the natural components of their environment into food, energy, and by-products that other creatures rely on. The plants of Hell Creek—from the cycads to the palm trees—took in carbon dioxide from the air and water from the soil. Sunlight hitting the green parts of the plants provided the critical energy source for water to be converted into oxygen and carbon dioxide to be made into the sugar glucose, the energy plants thrive on. No matter what kind of plant, from fern to flowering tree, each and every one relies on photosynthesis to grow and survive.

But now there isn't as much solar energy reaching the leaves. Some of the plants that are trying to grow seem sickly and brownish rather than vibrant green. Photosynthesis has not ground to a total halt, but there's just not enough energy for many plants to carry on as they had before. The adult plants were turned to ash, and the shoots can't find the energy they need, making this world just as harsh for plants as for animals. Of more than 130 different plant species that once grew thick through Hell Creek, only about a quarter remain. No complete families were lost, just as dinosaurs and mammals survive, but some plants are gone forever. Perhaps the most noticeable absence is *Dryophyllum subfalcatum*. This great, shaggy tree belonged to the same family as walnut and hickory. In fact, it looked something like a low-branching hickory tree with long, ridged leaves. In the spring, long, fuzzy-looking clusters of flowers called catkins grew all over, waiting for the wind to carry their pollen. But the diminished sunlight isn't enough for the surviving seedlings that have tried to push out of the soil. The tree that defined the Hell Creek forest is gone forever.

The plants with the best chance are those that evolved to

tolerate low-light conditions, those often kept in the shade by taller trees of the Cretaceous world or that otherwise, by sheer luck, could get by on less. Low to the ground, a small, spindly branch pokes up from the soil. The scraggly plant doesn't look like all that much, just a central wisp of a trunk with a few splashes of needle-bearing branches, looking more like a well-used dinosaurian toothbrush than a tree. But this will be one of the survivors. It's a *Mesocyparis,* a Cretaceous tree that was around for millions of years before the asteroid strike and will persist for at least the next 6 million.

Mesocyparis are conifers, belonging to the same family that includes junipers and redwoods. They're hardy trees, and the adult plants create round seed cones held in pairs; they look a little like prehistoric earrings. These tiny orbs will be what saves the plant. Or, rather, they'll be the enticement to get other species to cultivate a future for *Mesocyparis.* If the little plant can just hold on until that point, its story will continue. The birds will see to that.

A strutting, feathered form shakes the little tree as it moves by, throwing a few beads of moisture onto the avian's back. This is no imposing dinosaur. It's a bird, or an avian dinosaur—one of the few survivors of the dinosaurian family. Even compared to some of the birds that shared the Late Cretaceous world, the bird is on the small side. The time of giants is over. This little fluttering thing is what remains of the great dinosaurian legacy, one of the few surviving terrible lizards. There's no common name for the bird: humans capable of giving it one won't appear for about another 66 million years. But this is a ground-dwelling bird. A beaked bird. A bird set for survival even though it could have no idea what saved it or what would allow its descendants to thrive through subsequent ages.

Just twelve months earlier, the bird was but a small part of a much larger world. Its existence was just a slice of a rich, complex ecosystem that was brimming with all manner of creatures. Its life required awareness and even cunning. An omnivore capable of nabbing insects on the wing as well as grinding down seeds inside its gizzard, the bird occupied a strange place in the Hell Creek hierarchy. The avian dinosaur hunted and ate other creatures, but was often at risk of being eaten itself—silent raptor claws piercing through its coat of banded feathers to the pliant flesh beneath. Feathered non-avian dinosaurs—rather than birds, the avian dinosaurs—still ruled. Birds were just another source of prey. The Cretaceous avians were more aerodynamically adept, certainly, but they had no real defenses to speak of. If they mistimed their flapping away from pouncing feet and snapping jaws, they would become little more than piles of bloody plucked feathers scattered onto the understory.

This dinosaur-eat-dinosaur tussle had been going on for tens of millions of years at that point. Birds were not simply waiting in the wings when the asteroid struck. They were part of the great flowering of dinosaur diversity, their roots anchored deep in the Jurassic.

So much of dinosaurian life is focused on the large and superlative. But its greatest success story started so small as to be nearly invisible against the background of towering herbivores and carnivorous giants. The critical time was over 80 million years from this Paleocene moment. This was the Late

Jurassic, one of the many peaks in reptilian greatness. In places like prehistoric Utah, hundreds of miles to the south of what would become Hell Creek, giant horned carnivores such as *Allosaurus* and *Ceratosaurus* dined on the surplus of meat provided by eighty-foot plant-vacuums like *Apatosaurus* and *Brachiosaurus*. The giants can only ever give us a partial story, of course. Any giant requires giant-sized helpings of food—vegetable or animal—and those munchables come from an entire ecosystem of smaller creatures. These more diminutive and easy-to-overlook animals form the basis upon which everything else rests. And during this time—far from Jurassic Utah and Montana, on a little island archipelago that would one day become part of Bavaria, Germany—the first bird took to the air. It wasn't a very good flier. Shaped like a miniaturized *Velociraptor*, this dinosaur was covered from head to toe in feathers while still retaining teeth, claws, and the long scaly tail of its ancestors. This was *Archaeopteryx*, the "ancient wing" that has stood as the fossil keystone for birds in our understanding. Even though this urvogel still retained signs of its ancestry, and couldn't do much more than awkwardly flutter along, it still shared critical traits that link it to all birds that would follow over the succeeding 150 million years.

Most of the earliest birds resembled *Archaeopteryx*. They were small, exquisitely feathered, and predatory. These were fliers best suited to munching through the carapaces of insects and the scaly hides of lizards. There were no giant birds. Those niches were totally taken up by the other non-avian dinosaurs, including the raptors and their relatives. Likewise, the flying pterosaurs already dominated the air. Cousins of dinosaurs, pterosaurs were the first vertebrates to achieve powered flight and did so millions of years before birds would do the same. Through luck and happenstance, birds were able to find their place in the air and maintain a clawhold in a world that already seemed full of reptilian wonders.

Some birds held on to their teeth. Even as avians began to fly, many of these birds—recognized as enantiornithines— kept an array of small, peglike teeth that were quite handy for catching moving prey. These feathered creatures remained on the carnivorous part of the ecological spectrum, evolving to nab prey from the forests and the seashore. One of the first toothed birds discovered by paleontologists, in fact, looked like a toothed loon—*Hesperornis* was clearly a diving bird that snacked on fish—while the flying *Ichthyornis* was like a seagull with a sinister smile. Still, not all birds kept their toothy smirks. Some lost their teeth entirely.

The evolutionary process had happened before, even among some non-avian dinosaurs. When dinosaurs went from a mostly carnivorous diet to one that was more vegetarian, they often evolved a larger number of smaller teeth. While a *Velociraptor* was a carnivore with large, serrated teeth, for example, its relative, *Saurornitholestes,* had a similar body but a greater number of smaller teeth in its mouth, identifying it as more of an omnivore. Lineages that continued along that scale sometimes lost many or all of their teeth, developing toothless beaks. Birds did the same, only in the air. The trade-off wasn't about saving weight for flight—having teeth or not having teeth is only marginal in weight-saving terms—but owed to two major biological changes that would end up being the saving grace of these flapping, fluttering dinosaurs.

For millions of years after their origin in the Triassic, teeth helped set the timing of a dinosaur's emergence from its egg. That's because teeth take a long time to form, months of daily growth, so that baby dinosaurs crack out of their eggs ready to eat whatever they please. Even the most doting dinosaur parents were still reptiles and, as such, did not have any milk or other nutritious substance to offer their hatchlings in the earliest days of their lives. In fact, many dinosaurs didn't care for their offspring at all, leaving the little ones to fend for themselves and find what food they could. And the pressure was on to eat. No matter how large a dinosaur might grow to be as an adult, it started off life as small and vulnerable. The first year of any dinosaur's life was a parade of horrors, from drought to hungry carnivores ever ready to sup on the naïve. For many species, the only escape was to grow, and grow fast—a biological urge that required a great deal of high-energy, nutritious food. Baby dinosaurs had to be ready for this ecological struggle from the time they tumbled out of their shelly enclosures. That need required months of development, as many as three to six months for baby dinosaurs to be ready to face the world.

But without the need for teeth, embryos can develop much, much faster. The developing babies don't have to wait to sink their teeth into the world. During the Cretaceous, birds and non-avian dinosaurs took this route—an option allowed by the evolution of anatomical wonders very different from the crushing jaws of a *T. rex* or the grinding tooth batteries of *Edmontosaurus*. The origin of a gizzard opened up possibilities for dinosaurs that were not available earlier in their history.

Unlike the hadrosaurs, which evolved their own way of chewing, most dinosaurs could not chew. Their jaws worked more or less like shears, slicing or crushing or plucking food that was then swallowed whole. It's a quick way to feed, surely, but it creates problems when that food slides through the digestive system. Consider a carnivorous dinosaur, like *T. rex*. If this dinosaur tears off a large hunk of *Edmontosaurus* meat and swallows that chunk whole, the large bolus of meat is going to take longer to break down than if that dino steak had been cut up into much smaller pieces. It comes down to surface area—a hundred-pound chunk of meat has less surface area for its volume and will take longer to digest than a hundred pounds of smaller slices that expose a greater surface area to the digestive organs. But through variation and natural selection, some dinosaurs evolved specialized organs that could effectively chew food on the inside of their bodies—muscular gizzards that could grind down and break up food after it had been swallowed.

The ancestors of beaked birds—as well as some non-avian dinosaurs, such as the parrot-like oviraptorosaurs—evolved gizzards. That opened up the possibility of beaks. These dinosaurs did not need teeth to catch, kill, or break down their meals. They could focus on small prey like insects, lizards, mammals, and plants, plucking morsels to their reptilian heart's content and swallowing that food whole. Those meals would then be ground down by the gizzard, allowing the animals to draw in more nutrition, faster, than other dinosaurs. (Other species, such as herbivorous dinosaurs, increased the amount of time food stayed in their digestive systems by means of hindgut fermentation similar to today's elephants. Predators like *T. rex* simply filled their bellies and let the food run through quickly, effectively wasting food that could have fueled them and thus needing to eat greater quantities to keep up with the inefficiency.) And some birds even evolved another anatomical bonus—a crop. Located in the throat, the crop is a storage pouch. Cretaceous birds with this little pocket could pluck up their meals and store some, meaning that they could have a little bit of food already in their bodies, ready to be sent on to the stomach, even in times when food was scarce.

These important changes happened against the background of the dinosaurian heyday, while meek little birds were as much a part of any ecosystem as a long-necked sauropod or a multi-horned ceratopsian. The only reason they ever stood out as special is because some of these dinosaurs survived the worst of the K-Pg extinction. Nevertheless, being toothless was not a surefire ticket to survival. Beaks were not a surefire ticket to the Paleocene; some toothed dinosaurs—like little *Acheroraptor*—were able to survive beyond the terrible heat pulse simply by taking refuge in abandoned burrows. The underground world of Hell Creek had always been just as important as life above the surface, with mammals, lizards, and even a few birds digging down to create homes. Some of these dens became repurposed over time, a way for even more species to sleep in peace or escape the midday heat, and now there was more of an impetus than ever for Hell Creek's smaller inhabitants to seek whatever shelter they could find. This was life or death. Some *Acheroraptor* took refuge in them, as did various beaked birds. It was the chilly impact winter when toothed versus toothless became more important. Exposed to the relentless, hellish heat, beaked birds didn't fare any better than their other dinosaurian relations. The importance of beaks became clear only in the aftermath, as the survivors were left to pick and sort through whatever they could find in the ruin of the Cretaceous world.

Our hungry little forest dweller is doing just that as it silently steps through the ferns and ash, searching for a morsel.

A seed. A few flicks of the bird's toothless beak brings it up. To a plant, the seed is hope for the future. It is an investment, rich in energy that is ready to begin its transformation when conditions are just right. Not all seeds are hardy. Some perish quite readily. But others—by luck—had evolved to withstand wildfires and other catastrophes. These storage containers create a natural seed bank, ready to grow with the return of the light. But the bird has found this one first. Both are in an improvised evolutionary dance in which one relies on the other to survive.

Beaked birds are the only feathered creatures in this forest now. Even though birds, as a group, survived through this first harsh year, there have still been species losses as the avians pick through what's left to eat. The toothed birds, which had proliferated since the time of *Archaeopteryx*, are now entirely missing. Soon there won't be any more glints of tooth behind feather-lined jaws. There can't be.

Maintaining a mouth of sharp teeth comes with a reliance on animal food. Toothed birds ate other creatures—dragonflies, beetles, fish, mammals, other birds. Their diets were confined by their teeth. But when the aftermath of the asteroid caused global ecosystems to collapse, the destruction went nearly all the way down. Dragonflies eat other insects, but they are also food. Fish require sustenance, whether in the form of other fish, insects, or plants. A consumer that feeds on other consumers has very little to survive on now. But beaked birds do not face the same constraints. These birds nab arthropods when they can, but they are also capable of eating fruits, nuts, seeds, and the great smorgasbord of food that plants provide. The rise of flowering plants during the Early Cretaceous—about 60 million years before the Paleocene—led to an explosion of new, tough resources that allowed plants to reproduce in new ways, forming interrelationships with other organisms. Invertebrates began to pollinate flowers, which then produced fruits, which contained seeds that were eaten by animals and then deposited in new places by dung. In this slow-churning evolutionary two-step, some birds evolved the capability to break down the seeds themselves. They swallowed tiny stones that created grinding mills within their own bodies. Swallowed seeds weren't just jettisoned. The seeds themselves became food, all that potential energy diverted to the bird's survival and growth.

This relationship will benefit plants as much as birds. The surviving *Mesocyparis* trees, for instance, will eventually present their little seed cone baubles to the forest as they grow older. Birds, hungry for whatever resources they can scrounge, will pluck and swallow them, just as they did before the catastrophe. Now, however, the biggest seed cones are the most attractive—more energy for less work. The seeds have already adapted to survive the journey through avian guts, the trade-off being a chance to grow somewhere new, with a dab of natural fertilizer to boot. Repeated over and over again, birds will drive generations of *Mesocyparis* to present larger seed cones—twice the size of what they were at the end of the Cretaceous—in order to keep birds doing the trees' dirty work.

In time, the push and pull of this evolutionary relationship will lead to birds that are even better suited to finding and breaking open the seeds and nuts that plants so generously provide. There is no such thing as a one-size-fits-all beak, after all, and each has its own advantages and disadvantages when it comes to acquiring food. A beak that is good for cracking nuts is not the same as one that is good for drinking nectar. A beak that's good for catching insects will not be good for grazing on the ground. Depending on what birds do and what they eat, natural selection tends to nudge birds toward particular beak shapes and away from others. A long beak may be better than a short, powerful one for quick, snatching movements, so within a population of insect-eating birds, those with the biomechanically better beak shape will ostensibly catch more meals and have more energy to mate and rear offspring and pass that trait on. Do that over and over again and you can get transcendent change, with one species starting to look like something very different. In the case of functions like these, the beak might even reach an adaptive peak, or what might be thought of as the optimal shape. Beaks that just aren't quite the right shape will result in those birds having fewer offspring, with those lines possibly ending, until there's something else those birds do—a new food source, the disappearance of preferred prey— that gives the population a nudge in a different direction.

Often, these sorts of shifts happen early on in a group's history. Lineages very quickly spin off all sorts of forms, which then get tweaked this way and that over time. That's why the evolutionary tree of dinosaurs themselves, for instance, divided into three major groups very early on and, out of competition with crocs, spun off large carnivores, armored dinosaurs, long-necked giants, and more by the beginning of the Jurassic. The evolution of birds was similar—their disparity didn't pick up right away, during the Late Jurassic or Early Cretaceous. It didn't really start until the end of the Cretaceous and into the Paleocene. Prior to that point, beaked birds were competing with toothed birds, non-avian dinosaurs, and the flying ptero-saurs for niches. The world was already tightly packed with competitors, meaning that there weren't all that many niches open for bird evolution to fully take off. This similarity, in fact, might have made them prone to the mass extinction—a lot of birds doing relatively similar things through the Mesozoic. But with their competitors removed, birds now have an open opportunity. There are some things they can't do, like break bones or catch prey with claws or develop body armor. But past evolutionary changes have allowed birds to reproduce rapidly and grow fast in a world seemingly devoid of competition.

During the earliest years of the Paleocene, searching for nuts and seeds becomes the main way to get through the im-pact winter. During the lean years, birds with beaks best suited to plucking and cracking those foods fare best. But as the skies begin to grow lighter again, and more sorts of food start to become available—insects, lizards, small mammals, fruits—birds start selecting whole new ranges of food and undergo an evolutionary radiation unlike any seen before.

It's not just the birds that are awaiting the revival of the plants, though. For insects, a world without a forest is incredibly desolate. Not only are the non-avian dinosaurs gone, depriv-ing parasitic arthropods of their food and shelter, but the leaves that were so critical to many insects have yet to regrow. Insects that used to dig into leaves to lay eggs or winnow little path-ways along the green surfaces have virtually nothing left. Hell Creek's plants truly were the foundation of an entire ecosystem, and their vast die-off requires survivors to either adapt or perish.

Most insects seem to take the changes in stride, though. Even though some groups take a hit—such as springtails, aphids, and scarab beetles—there isn't a dramatic mass extinction of insects and arthropods. That doesn't mean that no arthropods died in the great conflagration and during the enduring months of impact winter. Untold numbers certainly did. But enough insects from the varied species and families did manage to hang on through the harsh conditions. After all, insects typically live short lives, leave copious amounts of offspring, and live at such a scale that a single patch of forest or meadow is like an entire world to them.

Still, the situation isn't exactly rosy now that many plants no longer grow. Insects that evolved in tandem with plants, using the great swaths of greenery for both food and shelter, have little to live on now. On the branch of a charred, dead tree, a small beetle does an awkward shuffle. Each of its segmented legs turn outward, giving the insect a little wobble as it moves—the domed shell on its back adding to its tortoise-like appearance. It's a lucky bug. Just months before, as a larva, the insect was fortunate enough to hatch onto a ginger leaf that had managed to persist through all the destructive chaos. The little larva mined its way through the leaf tissue, creating a winding pathway through the green cuticle.

Not just any leaf would do. The miniature world of a plant leaf—down to the biochemical components and defenses of the plant—can make a world of difference to a developing larva. Like other beetles and flies that wind their way through leaves during the early days of their lives, this particular beetle larva didn't roll around over the surface of the leaf but made its home inside, in the layers of the leaf that contained lower levels of cellulose—the tough material that helps plant cell walls stay rigid. Some leaf-mining species even evolved a specific relationship with plants that have chemical defenses like tannins, a natural astringent that helps ward off other insect predators while the insect larva stays safe inside. And it's this specialization that would make life much harder for insects that had relied on specific species of plant during the earliest parts of the Paleocene. While generalist leaf miners could deposit their eggs on just about any available greenery, specialized species—like the little tortoise-backed beetle—were not as evolutionarily flexible. When their host plants perished, so did the insect species that relied on them. Of all the insect species that were eating away at leaves in one form or another, about 70 percent have either gone extinct or been severely cut back by the collapse of the forests. The parasites and predators that relied on the herbivores are also not as abundant as they were just a year before. An extinction crisis is not simply a matter of survival or death. Even those species that manage to survive might face a dramatically altered world where finding food, mates, and shelter is incredibly more difficult than it was before.

One of the few holdouts is a small caterpillar, hugging the edge of a sickly leaf as if its life depends on it. One mouthful at a time, the green larva takes a bite of the leaf's edge and mashes it up, fuel for its long slumber before emerging utterly transformed. Lepidopterans—the family of insects that includes moths and butterflies—have been doing this for over 200 million years, very hungry caterpillars stuffing themselves on Mesozoic salads before using all that leafy energy to metamorphose. And, at the family scale, they'll survive. There will be just enough leaves to carry the population through the next two years of diminished sunlight, an eternity in insect terms but not so long as to drive the entire clade extinct. Unfortunately, there's more to survival than just finding food. The verdant color of the caterpillar, normally matched to the green of the Hell Creek forest, stands out just a little too much in this struggling habitat. A flash of feathers and the stab of a beak, and the caterpillar—leaf-stuffed—slides down the gullet of a foraging bird, just that much more energy to keep the hot-blooded dinosaur alive.

After the heat, the fires, and the cessation of sunlight, seeds are among the only accessible foods left. Beaked birds, or at least those adapted to breaking down tough foods inside their bodies, are able to seek out and persist on the great seed bank that was left behind by the charred forests. Each seed presents possibility, and, ingested, make hope the thing with feathers.

Elsewhere
THE PREHISTORIC NEW JERSEY COASTLINE

The crocodile swims over a submerged graveyard of bones.

The skeletons are not prey. Even at twenty feet long, the *Thoracosaurus* isn't voracious enough to have consumed all the creatures whose scattered bones poke from the sand. The motion of the waves rocks them gently, giving the dead motion even a year after their demise.

Our *Thoracosaurus* is one of the survivors. Just hundreds of days prior, this reptile had stiff competition in the nearshore waters. Massive, serrated-toothed sharks munched on dead dinosaurs that found their way into coastal rivers, decomposition gases bloating the bodies of the hadrosaurs and tyrannosaurs until they arrived just offshore. The sharks were all too glad to pop the stinky balloons and feast on whatever tatters they could carve off. But most of the time these predators ignored the *Thoracosaurus*. She had grown too large to be nabbed as a snack, even if one of the sharks had made a meal of her left foot when she was younger. No, far worse was *Mosasaurus maximus*.

No creature here grew as large as *Mosasaurus*, up to forty feet of voracious, seagoing lizard. Much like *Thoracosaurus* herself, the large mosasaurs were ambush hunters, taking lungfuls of air and sinking just deep enough that the mottled gray of their backs began to blend into the darker shades of the waters below. Often, strange shadows suddenly zoomed close to fill a victim's vision with pink triangles of tongue and teeth—with a secondary row inside, on the roof of the mouth, situated to prevent any wriggling prey from fortuitously flopping themselves out of the mosasaur's mouth.

But such big bodies required a great deal of food. After the heat, when the skies grew dark, the ocean began to grow deathly quiet. The ammonites and belemnites disappeared. So did the large sea turtles that bobbed along at their sluggish pace, protected from predation by armor but unable to find enough food to sustain themselves in the post-impact waters. Fish became rare, too, as they struggled to find enough plankton to feed on. The *Thoracosaurus*, with her great snaggle-toothed smile of thumb-sized teeth, was the last guardian of this place, welcoming unwary visitors into her gullet. Soon, even her mighty osteoderms will become little more than a scattering of scutes on the sand.

7

One Hundred Years
After Impact

Three years of darkness. For three entire years, as Earth continued spinning along in its orbit, the sun's rays could do little more than caress the planet's surface. Day after day after day passed under a blanket of debris and dust.

From the perspective of deep time, the millions and billions of years of Earth's existence, three years might seem like a trifle. The great and now vanquished *Tyrannosaurus rex* had a tenure of 2 million years, long enough for even this rare, large carnivore to develop populations in the billions. Three years is just a fraction of a fraction of that one comparatively short dinosaurian span. But this is not a matter of all-time records or chronological extremes. The impact winter was enough—enough to drive photosynthesis to a near-halt and cause the rapid collapse of ecosystems on land and sea. In fact, ocean creatures suffered even more than the terrestrial organisms. The ocean's ecosystem is based on photosynthetic plankton. With planktonic populations eroded by the diminished sunlight, the very foundation of ocean ecologies was destroyed. The ecological carpet had been ripped out from beneath the complex and interwoven food webs that had evolved over tens of millions of years. For a time, the basis of the seas was shifted to plankton that could hunt and consume and proliferate, taking over just rapidly enough to prevent total extinction. Without their predatory habits, the oceans may have very well returned to a single-celled state that the world had not seen in over half a billion years.

The basic conditions of the planet—the climate, the amount of sunlight reaching the surface, the patterns of storms—dramatically changed. There was no slow creep of alterations that multicellular creatures could adapt to. The changes played out within the lifetimes of many organisms, almost like a switch being flipped. Among the alterations was something that had only rarely and temporarily affected life on Earth—acid rain.

The sulfates thrown high into the stratosphere did more than reflect back the nourishing sunlight that had been taken as a given by Earth's organisms for so many millions of years. The aerosolized sulfur from the target rock often combined with oxygen in such a way to create sulfur trioxide. Strewn about by air currents, the compounds remained in the air as rain clouds formed. Typical rain has a pH of about 5.6, a fact that life on Earth had become accustomed to during its long evolution. But this rain contains sulfuric acid, enough to drop the pH of rain to about 4.3.

The rain doesn't melt organisms on contact or hiss and sputter when it hits the ground. In fact, acid rain isn't even as acidic as lemon juice. But it's enough of a drop in pH that lakes, ponds, and oceans become more acidic as they collect the altered raindrops. Organisms that rely on shells for protection cannot build them as reliably. And on land, plant life becomes discolored—turning sick shades of rust brown—as chemical reactions within the soil eat away at the nutrients trees and other vegetation need to grow. It's a slow, nearly invisible process, subtly altering the conditions of life that Earth has become accustomed to. The rain took the sulfates out of the atmosphere, offering some hope that the sunlight might return sooner, but at a terrible price to the world below.

In Hell Creek, there were few trees to be affected by the acidic downpour. During the years of impact winter, when acid rain became another tragic fact of life, much of the surviving plant life waited belowground. Seeds and nuts were the hope for the future; fast-growing ferns and other plants poked up here and there, while mushrooms and other fungal colonies grew from the death distributed by the impact. Every patch of ground was a fraught space, new life desperate to take hold in the charred remnants of the single worst day the planet ever experienced. The struggle flowed through Hell Creek's waterways, too.

In the hours and days after impact, the ponds, lakes, streams, rivers, and marshes of Hell Creek were a refuge. The creatures that lived there were almost unaffected by the intense, violent heat. But now the environments that saved the fish, reptiles, and amphibians have become traps. The bodies of water collect the showers of acid rain, turning more acidic themselves.

On the edge of a deep pond, a depression created by the repeated footfalls of dinosaurs moving over soft ground, there rests a little frog named *Eopelobates*. The amphibian wouldn't look out of place in a pond of almost any age. From its rounded snout and shield-shaped head to its long, spindly legs, the frog is the epitome of its semi-aquatic family. And for such a creature, which relies on water to hunt, mate, and breathe, acid rain would seem to be an insurmountable problem.

Some lakes and ponds seem to be less lively after the showers of acid rain. The rapid shift in pH is too much, disproportionately affecting the soft eggs of spawning amphibians and newly hatched tadpoles. It's a disaster for sensitive species. For the most part, amphibians play the numbers game when it comes to reproduction—just as the non-avian dinosaurs had. Instead of investing incredible amounts of energy in watching over the next generation until their offspring are large enough to fend for themselves, amphibians lay dozens, hundreds, even thousands of soft, globby eggs that, if fortunate, will not be eaten by fish, turtles, or other amphibians. From a starting point of thousands in a single clutch, hundreds may become tadpoles and scores may survive to adulthood—a fraction of the starting number but enough to survive. But the acid rain affects some species in their most vulnerable moments. Many eggs don't hatch. Many tadpoles don't thrive. Even though not every species is affected, there is little escape for those that are. Amphibians are inextricably tied to the water, and their advantage after the initial impact has turned into a terrible liability.

Geology, however, sometimes comes to the rescue. The chemical signatures of ponds and lakes are affected by the soils and basins the water fills. No two are precisely the same. And in Hell Creek, many of the sheltering waterways are carved into limestones and soils that contain the crumbled components of these rocks.

Strange as it may seem, Hell Creek already has a fossil record below it. Millions of years before the impact, millions of years before the origin of *T. rex* and *T. horridus,* this area was covered by a vast shallow sea. Clams, mussels, oysters, and other shelly animals thrived along the bottoms, providing so much food that some marine reptiles—like the mosasaur *Globidens*—evolved blunt, dome-like teeth to crush the shelly bounty to bits. In time, however, the seaway began to drain off North America. The waters became shallower and shallower, the great stacks of shells and shell parts transformed into limestone.

The shells of the bivalves and mollusks that lived on the ancient seafloor were principally made of calcium carbonate, $CaCO_3$. Calcium carbonate reacts with sulfuric acid, the chemical dance between the two buffering the effects of the acid and neutralizing it. For living animals, that can be bad news. It's hard to make a home out of shell when that shelly material is being eroded as you make it. But the shells in the limestone are from creatures long dead, and so the fossils are a boon to the Paleocene animals that are still hanging on. The remnants of the fossilized, lithified dead are the saving grace of Hell Creek's amphibious species. In ponds and lakes formed in soils and rock exposures rich in $CaCO_3$, the ground itself is enough to counteract the acid.

Eopelobates is lucky enough to live in such a pond. The waters here haven't changed much since before the impact. The more immediate concern, within the boundaries of the lake, is the cold. Frogs are ectotherms: their body temperature is regulated by the temperature of the surrounding air and water. They're more active during the hot months, but their physiology requires that they slow down when the temperature drops— just as it has during the impact winter. It's getting too chilly to effectively chase after prey. Every attempt the frog makes, trying to dart fast enough to nab the equally nimble tiny insects, is becoming a little slower, a little more difficult. *Eopelobates* is expending more energy than he's taking in, a path to starvation if the frog keeps it up.

With a kick, a push, and a few strokes of his back legs, as he swims *Eopelobates* swims down the subaqueous incline of the pond toward the muck and rotting vegetation of the bottom waters. This will be his refuge, at least for a time. While his temperature-dependent metabolism is a liability when he's trying to hunt in cooler times, it also provides him with an ability not enjoyed by the warm-blooded endotherms. *Eopelobates* can wait. If he can find a quiet patch of brown sediment at the bottom of the pond, he can burrow in and wait for a while. Perhaps for a long while.

The frog can't just pick any old spot and burrow deep down, like a turtle does. There are already some shelly reptiles down below the surface, nestled into the mire and extracting what little oxygen they require through their mouths and cloacae, intending to sleep through the seemingly endless winter. Being completely encased in mud doesn't matter to those reptiles as they've switched over to alternate ways of extracting oxygen by now. But *Eopelobates* can't do that. Frogs need to be in touch with the water. Even when they hibernate, slowing down and dozing through the cold, they still extract oxygen through their skin. Part of their body needs to be exposed to the water, enough to keep respirating despite being submerged. It's a strange physiological catch. Fish can breathe through gills. Turtles breathe with lungs, but they can shift gears to take in oxygen in other ways. Frogs are somewhere in the middle, making do with what millions of years of evolution has accidentally granted them.

After a few swim strokes, gliding over the bottom and trying to avoid the sharp jaws of any hungry turtles that are still awake, *Eopelobates* touches down on the bottom. Brown shreds of decaying vegetation float up and do a little slow-motion twirl as the frog remains still for a moment. Then, first with one leg and then with the other, the frog starts to shuffle backward with his feet. Inch by inch, he backs just far enough into the sediment, leaving just the top of his back, his eyes, and his snout exposed. He can stay like this for quite a while. Weeks, perhaps months, will pass by as he waits. Now and then he'll stir, taking a very slow lap around the bottom of the pond, but he'll always settle back in, at least until temperatures rise just enough that the aquatic insects and other morsels start to get more active again. This is a slowdown, after all, and not a cessation. His body still needs energy, still needs sustenance. This is a time to stretch each one of those crunchy calories, perhaps long enough to see the sun's full warmth return. Here, cradled by the limestone of an ancient sea, he can wait.

Not every place is so sheltered. There is little escape from the acid rain for organisms that live in the soil. Worms can move, but bacteria, fungi, plants, algae, lichens, and other forms of life can only grow, die back, and grow as best they can in response to the changing conditions. And it's within these soils that the record of this extreme disaster has come to rest.

The geologic hallmarks of the environmental catastrophe are already being scattered over Hell Creek, locked into sediment that will become sandstone, mudstone, and other sedimentary rocks. It's an invisible accumulation, unfolding constantly but involving particles too small to notice as the minuscule tidbits settle into places where sediment is being laid down. That's a critical part of the record. If Hell Creek were a mountain habitat, nothing at all would be left. Topographically dramatic as it might be, a mountain is an erosional environment. It is a place where water, wind, sunlight, and other natural forces are winnowing away at the towering stone, pieces constantly carried downhill to the lowlands. The animals that inhabited high altitudes during this era will not leave any legacy behind. They will be erased, save for bones ground down to specks and carried down rivers and mudslides to lowlands. But Hell Creek is still a wet lowland, a damp, relatively flat place where the various waterways help to coat and cover parts of this world, including direct products of the impact.

The asteroid is not sitting near the equator intact like some terrible monolith. Its collision with Earth pulverized it just as the impact destroyed vast amounts of the target rock. Pieces of the asteroid, rich in the metal iridium, started falling back to Earth in the hours and days following the impact, becoming locked in clay and other sediments. The glass spherules, shocked quartz, and other debris from the strike point have been preserved here, too, a record of the violence of those first moments. The soot and carbon from the forest fires were worked into the mix as well, a testimony to the great conflagration that ripped through the forests that once grew tall enough to shelter stalking *Tyrannosaurus*.

Bones of the fallen, though, are not quite as durable. Bone is a strange tissue. Consisting of a combination of flexible collagen and rigid, friable hydroxyapatite, bone itself—the biological compound that makes up each element that fits into the skeleton—is a durable, flexible scaffolding. The ancient, outer armor of twitchy fish in ancient seas became modified to be an internal support system capable of supplying the body with minerals when needed, repairing breaks, and growing to transform embryonic *Triceratops* that fit in eggs the size of grapefruit to giants over thirty feet long. Along with hard, enamel-covered teeth, bones last far longer than the rest of the body.

In the years since impact, the flesh of the deceased Hell Creek animals has decayed away. Skin, fascia, muscle, viscera: all gone to reveal the supportive structure beneath. The exposed bones are eaten and burrowed into by other creatures, drying out and losing the marrow, blood vessels, and nerves that once grew within the skeletons in life. And if left otherwise undisturbed, the bones start to go through a long process of weathering.

A newly denuded bone is smooth. It represents what the

structure looked like inside the body, the tissue intact and not warped by exposure. Over time, however, bones begin to change. They become more brittle as the protein collagen rots away, leaving behind only the hard, mineral part of the bone. Over time, the tissue warps. Tiny cracks start to run over the surface—the longer a bone has been exposed, the more cracks will proliferate and the more surface area will be exposed to the very forces that are destroying the bone. Wait long enough and the bone will turn into dust.

The vulnerable remains of the dinosaurs, pterosaurs, mammals, and other impact casualties won't last even that long. They're not going to crumble. They're going to dissolve into nothing, as if the aftermath of the impact is trying to erase the horror of what transpired. Acid rain, by direct contact or by seeping into soils that have begun to cover Cretaceous remains, eats through the bones. Over time, as the sulfur is taken out of the air by rain and brought down to the surface of the planet, the dead are being etched away to nothing.

There should be great boneyards from this time. Waterways help collect and entomb the dead for millions of years in this place. The rocks should contain a vast killing field, the majority of the Cretaceous world suddenly expired in one great graveyard. But that's not to be. The record, as it's coming together, will seem to peter out toward the time of impact. The direct record of the years immediately following will largely be lost, leaving behind only clues that were sturdy, durable, and fortunate enough to persist in the rock record. Even the recently buried dead, those covered up in the years leading up to the impact, will be lost. As the water leaches into the ground, it seeps into bones that had already won the tapho-nomic lottery and been buried. The remains will be broken down little by little, seemingly leaving no trace of the animals that lived in those last Cretaceous days. As the fossil record takes shape over the coming years and millennia, sediment will turn to stone and lock what's left into the annals of deep time. In the several feet below the boundary layer, the record of the great dinosaurs will largely be eaten away by acid, the rare *Triceratops* skull attesting that dinosaurs really did make it to the very end.

Elsewhere
WESTERN NEW ZEALAND

A beam of sunlight breaks through the rain clouds of another wet, soppy day. Ferns, holding droplets of water along their leaves as if they were precious jewels, are the recipients of the sudden warmth, as are the mushrooms scattered through the field. This is what a meadow looks like in the early days of the Paleocene—not a place of grasses or wildflowers hemmed in by trees but a vast, shivering field of ferns and fungi.

There are a few charred stumps and spindles of dead trees here and there. Most have now been returned to the soil, their carbon-rich bodies enriching the ground for the forest's great comeback. But that will still take some time. Despite being a world away from the site of impact, this place was cut deep.

On the last day of the Cretaceous, this forest was already something of a relic. Even as flowering trees were overtaking the Northern Hemisphere, providing bees and other pollinators with plenty to do, this place was a great conifer forest that provided as many needles as the great long-necked sauropods could eat and offered concealment to great, sneaky carnivores with immense, rending claws on their hands. Armored giants snuffled along, weaving their way through the stands of fibrous trees, annoying the beaky, two-legged herbivores that gamboled through the understory. That world was lost to heat and fire, then dark and cold, the immobile plants just as susceptible to the terrors as the great dinosaurs were.

But the great variety of life is incredibly difficult to extinguish. Variations between individuals or species offer just enough luck to edge by, to set a new beginning. Fungal root systems threaded deep enough in the soil to escape the worst of the heat. The seeds of ferns were already planted, ready to sprout. Together, they took the black and white of an ashen landscape and returned it to its warmer, richer earth tones.

To see so many ferns in a place ravaged by fire a century before might seem strange. But after the curtain of debris finally lifted from Earth, and weather patterns began anew, there was just enough wet here, just enough damp and acidity to the soil to favor the ferns. They didn't simply grow—they flourished. A living carpet shared with fungus, what would have been a great smorgasbord for the shuffling, snuffling dinosaurs whose bones were built upon such plants and then returned to the dirt. Birds and a few fortunate mammals small enough to hide beneath the fronds are the only creatures large enough to easily spot here now, but there will be more. On the edge of each little droplet of rain, the sunlight creates a small glimmer.

8

One Thousand Years
After Impact

The day is already getting too hot for the little snake. With the sun rising high on the way to noon, the tiny squiggle of a serpent—only three inches long—winds through the forest of ferns toward a shady shelter to while away the hottest part of the day, content to wait for the cool of the evening to hunt again.

Life isn't easy for an ectotherm. Compared with the great dinosaurs that had roamed this place a thousand years before, or the mammals that could slurp up the little snake like a scaly noodle, tiny *Coniophis* seems to live a slower, more retiring existence, following the rhythms of the surrounding temperature.

Being cold-blooded served *Coniophis* well during the times of terror and cold following the asteroid strike. Being small and prone to take shelter underground, the snakes made it through the incredible post-impact heat in underground dens created by other animals. But the long winter that soon settled in was another problem altogether. Temperatures got too cold for the snakes to stay active. The air was chilly and sunlight was weakened by the dust cloud swirling in Earth's atmosphere. It was just warm enough underground for the snakes to forage for invertebrates, find mates, and survive, but life had slowed down in the chill. That might have saved the snakes.

If *Coniophis* snakes had been endothermic—warmed by the internal processes of their body—they would have faced a big problem during the post-impact winter years. A hot-running body demands to be fed. Maintaining warmth is physiologically expensive, and just as a fire needs a great deal of fuel, so does a warm-blooded creature. Had this been the case, the snakes might have starved. Aboveground invertebrates and other morsels were severely cut back by the intense heat and the lingering cold, and endotherms were challenged to find enough food to stay alive. But ectotherms play by different rules. Just as frogs can slow down their metabolisms and go into a torpor, and just as some turtles can slumber in the pond muck for long periods of time, snakes can go into their own form of hibernation—a slowing down in cooler times called *brumation*. *Coniophis* and other surviving snakes didn't cease activity altogether: it wouldn't have been possible for any of them to last three full years without food. But it's a biological change in tempo, a response to the environment that lets their bodies subsist on less, making every meal count just that much more. It's a gift to the species that don't live life at such a frenetic, warm-blooded pace. After all, the dinosaurs were diverse, disparate, extremely active, and always, always hungry. Under the stress of the catastrophe, all that was left of them were small, feathery creatures that poked around in the dirt for whatever seeds and insects they could find.

Ready to escape the heat—and the pecking beaks of the avian dinosaurs—little *Coniophis* slides down into the cool and dark of a mammal burrow, slipping off into an abandoned side chamber to rest. There is no impenetrable fortress in the Paleocene world, no place where any creature is fundamentally immune to becoming a meal for another, but it's quiet and safe enough. It'll do. And it's what snakes have been doing since the time of their origin.

Even by the time the vast asteroid struck the Earth's surface, snakes were survivors. The first serpents slithered through the fern-covered floodplains of the Jurassic over 100 million years earlier. At a glance, they didn't look much different from a slender lizard. That's because snakes evolved from lizards, becoming longer and moving on their bellies instead of with their legs. It'd be millions of years before limbless serpents started to slink over the planet's surface—back in the Jurassic, the earliest snakes still had tiny, twiddly limbs sticking out from either side of their bodies. And they burrowed.

Swimming through soil is no easy task. Small body size helped the reptiles find pathways through the earth, but legs become something of a liability while doing so. On the surface, limbs can push reptiles along. In the water, limbs can paddle. But belowground, among the packed particles of soil, legs had to be tucked alongside to keep out of the way. There wasn't a way to bring them forward to do a full stroke, so early snakes wiggled their way through instead. The shift fundamentally changed their bodies, down to their inner ears. Sound is quickly dampened by soil, but snakes still need to listen to the world around them. Over time, natural selection adapted their ears to be more sensitive to the low-frequency vibrations that travel farther through soil. Little *Coniophis* can slither along as easily on the surface as underground, but the reptile's afternoon snooze in a burrow is a whisper of his ancient, scaly ancestry.

The tiny serpent is a Cretaceous leftover, now facing a world that seems full of possibility. The great, towering dinosaurs are gone. Flowering plants are starting to come up where conifers once dominated. These creatures of the freshwater realm bask in their pools and watch the world begin to recover and re-seed. But there is no script for what's about to unfold, no cast of characters that inevitably must be filled. Life is just beginning to spring back, certainly, but it would be a mistake to think that this is simply a matter of healing. The concept of healing involves a return to how things were before, perhaps with old niches occupied by new species and familiar patterns of ecological complexity returning. But that's not how life works.

Consider the expanse of what was once was the Hell Creek ecosystem. This is no longer a land of bones and ash, but it's not the thrumming, buzzing habitat it once was. Life has gone back to a smaller scale, populated by survivor species. Flowering trees like dogwoods and magnolias are proliferating, with dawn redwoods representing the coniferous contingent. The remaining mammals are mostly burrowing species—the great menagerie of odd Mesozoic mammals has largely been pared back. Insects still call and chew and swarm, but there are fewer species now. The freshwater environments experienced the least amount of change—the fish and frogs and turtles and champsosaurs still carrying on as they had at the end of the Cretaceous—but the biodiversity in this new world is a fraction of what once thrived here. It's going to take time for life to bring about more life.

Looked at one way, this landscape is full of empty niches. There are no megaherbivores here. All of the plant-eaters are small. They help disperse seeds, and undoubtedly affect the survival and evolution of the plants they eat, but none of the surviving mammals are ten-ton monsters that topple trees and trample lowlands into ponds. Nor are there any giant carnivores—an ecological impossibility when all the potential prey are snack size. The airborne giants are gone, too, as are all the toothed birds and other carnivorous fliers that nabbed insects and lizards in the forest canopy. The dramatic loss of so many species isn't simply a death tally. Each species is a break in the ecological web, multiple spokes left hanging and tumbling down. Nature would seem to be shot through with gaps.

But evolution doesn't require a gap to thrive. If anything, the existence of one species creates interactions and roles that often spurs the origin or proliferation of others. Diversity generates diversity, not ecological space alone. A mass extinction can help shake things up or provide opportunity, but it's the interaction between organisms that provides evolution's momentum.

There is no figurative landscape of available niches that organisms will inexorably be ushered toward to replay the same interactions time and again. There is no optimal state that the living world will return to given enough time. What's being woven now is something entirely new.

Throughout all the ages of life on Earth, through the annals of deep time, the eras of greatest biodiversity have been those when organisms transcend some barrier and begin to bump and nudge each other in new habitats. During the Carboniferous, over 290 million years earlier, vertebrates began to venture out of the water and onto land. Fish still thrived in the seas, but now—to take advantage of the insect-filled forests buzzing with crunchy meals—fish began to thrive on land. Some of these amphibious animals stayed tied to the water. Other lineages, with assistance from the newly evolved amniotic egg, began to make a home on land. On landscapes where there was once little more than rock and bacteria, towering forests thick with amphibians and arthropods grew. The change was so dramatic that bacteria had to catch up. Primitive plants evolved a new, hardy material—lignin—to grow into tall trees. Bacteria that could break down lignin had not yet evolved, and when these massive trees fell they decomposed so slowly that they were more likely to be buried in massive swamps, their carbon eventually turned to coal. This great burst of activity did not follow a mass extinction or play out as species filling prescribed niches. When plants rooted themselves to shore, they made new habitats for invertebrates, which in turn lured in vertebrates. Each evolutionary happenstance opened up new possibilities, biodiversity generating itself through interaction.

If there were a preordained path, or convergence were the

rule, then the Hell Creek survivors would be ushered toward all those roles that were vacated. In time, there might be towering, tyrannosaur-sized mammalian beasts or flying lizards the size of a pterosaur. After all, this would seem to be the big evolutionary chance for the survivors—especially the mammals that had long stayed small because of the dinosaurs. But there is no race, no rush, no urge to recapture what once was. Each action, each biological interaction and process, will slowly and subtly direct what's to come, the survivors creating niches rather than being beholden to them. There are no defined boxes that species must fit within. Natural selection, that persistent driver of change, requires variation to work on—from organisms' resistance to disease to the components of their diet and the shape of their bodies. It's an incredible, incalculable, intertwined set of small differences that allow ever more complex ecosystems to evolve. The emptiness of a post-extinction habitat offers a great many possibilities, certainly, but an ecological lack doesn't inherently spur ecological fullness. It will be up to the survivors, to imperceptibly shape the world from this new starting point.

Not all of the survivors will thrive again. Some of the species that were able to persist through the fiery hell of the first day and intense cold of the impact winter are shuffling toward extinction, not recovery.

A distant relative of little *Coniophis* rests under the shade of a young ginkgo tree, her scaly skin dappled by light filtered through the shell-shaped leaves. The length of the reptile's eight-foot-long body is largely gray, dotted here and there with tan spots, a perfect match for the quiet spot the monitor lizard has chosen for an afternoon respite. This is *Palaeosaniwa,* the largest lizard of Hell Creek. And she's among the last of her species.

Just a thousand years prior, *Palaeosaniwa* was the lizard equivalent of *T. rex.* Adults could grow to over nine feet in length, their jaws fitted with incredibly sharp, recurved teeth that made short work of mammals and baby dinosaurs. But in the hours after impact, most of these large lizards were unable to find suitable shelter belowground. They were too large for the burrows and had no instinct to create their own. Most perished. Only the smaller, immature lizards were able to find refuge. And these youngsters emerged into a world where the future of their species rested on them, and them alone.

Carnivores haven't had it easy in the post-impact world. Meals were scarce in the early years. *Palaeosaniwa* grabbed what they could among the surviving beasts, birds, and reptiles, raiding nests and burrows to sate their hunger. Because they were ectotherms like *Coniophis,* the lizards were able to maintain a toehold, but the population has never quite recovered. A thousand years after impact, just at the dawn of a new evolutionary flowering, the lizards are finding adequate prey but fewer and fewer mates. There is enough for a few to survive, but not enough to ensure a breeding population—especially since it takes years for these lizards to reach sexual maturity. Not very long from now, *Palaeosaniwa* will become functionally extinct.

Extinction doesn't occur when the very last member of a species takes its last breath. That moment is the conclusion of a life, but a species may have become functionally extinct well before then. It's the difference between proximate and ultimate causes. The last *Palaeosaniwa* may perish in the jaws of a predator, or die of a disease or old age, but whatever the reason, the species is already beyond the point of recovery. The incredible violence and chronic cold of the post-impact years winnowed the lizard's population down to a fraction of what it once was. Difficulty finding food in damaged, ashen landscapes allowed only a small portion of lizards to survive each year. Those individuals have since spread out over the landscape, requiring years to grow before they even attempt to find a mate, create a nest, and watch over their eggs, and none of those activities have any guarantee that a new generation will hatch and be successful. This is extinction by attrition: *Palaeosaniwa* is maintaining only a tenuous hold on existence. The lizard won't even notice its own extinction.

In time, extinction comes for all species. Some leave descendants. Others do not. Beautiful as the image is, there is no tree of life. The shape of biodiversity is more like a chaotic blanket, individual threads splitting, being snipped off, branching again, creating an incredible tangle of species that are both discrete and connected. All the species alive in this moment, at the dawn of the Paleogene, will eventually perish. But some will sprout populations a little different from their point of origin, variations that will survive even as their parent species disappear, and with them the same ecological dance will begin again. The species that exist today will shape what tomorrow looks like, life itself driving the profusion of so many unique forms.

In what remains of the Hell Creek, a new garden is just beginning to grow. The seeds were literally sown in the Cretaceous, a biological remainder that has set the foundation for the next 66 million years. *Palaeosaniwa* flicks her tongue, smelling the air of another Paleocene afternoon.

Elsewhere

THE NORTH ATLANTIC

The tiny sphere bobs along near the surface, the smallest of specks in a seemingly infinite sea. Up close, however, the dot is not just some oceanic dust mote. The aquatic smidgen is a sphere made of overlapping discs, donut-like plates with a small, whip-like flagellum sticking from each. The individual parts are coccolithophores, glommed together to make a coccolithosphere.

So close to the surface, bathed in sunlight, each part of the sphere is filled with a deep, rich green. The coccoliths are photosynthesizing, using sunlight to create their food. The sphere could simply bob along, day and night, the planktonic alga sustaining itself from the daytime rays. But the flagella are a sign of harder times, back when this particular form of coccolith was the dominant predator in these waters.

The seas almost died during the impact winter. Sunlight is an essential part of ocean ecology, allowing algae and other photosynthesizers to thrive. In the dark years, those photosynthesizers couldn't carry out their role—depriving the organisms that fed on them of food sources. Chip away at the foundation and the entire ecological edifice can come tumbling down. Many species of coccoliths died out, but not all. The survivors were able to switch to a different diet.

Evolution doesn't plan ahead. What's allowed life to survive through billions of years is variation and happenstance. The greater life's diversity, the more likely it is that there will be some individuals, some populations, some species that are able to cope with the stresses that push and pull at life. In the case of the coccoliths, some Cretaceous species were *mixotrophic*—able to absorb small bits of organic matter as well as make their own food. These were more mobile species, able to move back and forth by lashing their little organic whips, digesting smaller organisms and detritus when the sunlight dimmed.

The tiny algae kept the oceans alive. Even with the terrible planktonic crash, the mixotrophs proliferated and provided a bridge from the days of catastrophe to recovery. Some planktonic predators were able to subsist on the coccoliths, helping to keep parts of the ocean alive. And when the dust and sun-blocking debris finally wafted out of the atmosphere, the surviving coccoliths could take up new roles again. New photosynthesizers began to evolve from the mixotrophs, beginning to restore the algal web that's so critical to ocean health.

If all the Cretaceous coccoliths had been winnowed into perfect photosynthesizers, the impact winter would have wiped out the seas except for bacteria. But these organisms playing various roles allowed some to survive. It's not so much chance as consequence, the range of possibilities for life constantly shifting. Survivors aren't born so much as selected by happenstance.

The little ball of organic discs continues to float along in its microscopic world, a dense and thickly populated landscape in what might otherwise look like clear water. A few of the sphere's flagella twitch and move the ball to the side, engulfing an even tinier morsel from the water.

9

One Hundred Thousand Years After Impact

Lunch. The little orb weaver moves quickly. The small beetle is bucking and buzzing, flittering its wings to try to escape the stickiness of the web, and it just might. Spider silk is a strong material, but a web is made of many threads woven together. If an insect struggles hard, a meal might turn into a shredded tapestry and another night's work.

The orb weaver is fast. She quickly embraces the struggling beetle and puts her legs to work, spinning silk to cocoon and immobilize the insect. But then both spider and beetle are gone, a mash of legs and exoskeleton in the jaws of a twenty-pound mammal. This is *Baioconodon*, foraging for whatever she might be able to find in a world that has no memory of the extinction even as its scars are still visible.

Baioconodon is an odd little quadruped. Her toes are tipped in blunt little hooves, yet she has a fearsome smile. Her canines are proportionally long compared with her rough, crested cheek teeth. If she wanted to, she could easily sink these piercing teeth into other smaller mammals running around the forest. But she has no such inclination. She's an omnivore, munching on leaves and lizards, fruits and insects, and those pointed teeth are principally for display. Out of mating season—like it is now—she'll open her mouth wide, snarl, and hope to drive interlopers off. This is her little patch of Paleocene bounty to pick over as she pleases.

Ferns sway out of the way as she continues on through the forest. Enough time has passed that there are no longer visible carbon-covered trunks and logs from the great fires of the post-impact world. Trees have grown and fallen, and new ones have sprung up in their place, moss-covered trunks on the forest floor interspersed with a canopy that might, for the first time in a long time, be called old growth.

But the real success here is owned by the ferns. The fronds leave patches of damp on the fur of *Baioconodon* as she moves through the understory, beads of morning dew dappling her brown fur. Patches of fiddleheads are seemingly everywhere, almost like the heads of a hydra in repose. For any one fern that might perish for lack of sunlight or moisture, there are many more to take its place.

This is the height of the fern spike.

Ferns are among the most ancient plants on the planet. The first of these feather-like plants evolved about 360 million years ago, back when the ancestors of dinosaurs were just beginning to come ashore in the form of four-legged amphibians. Ferns evolved to thrive in wet, low-lying environments, forever tied to places where moisture is plentiful.

The life cycle of ferns dictate that they never stray too far from moist, squelching environments. That's because these understory plants grow from spores rather than seeds. Ferns do not wait, all tucked up inside a safe, armored shell, but instead have an alternating, back-and-forth life cycle in which the wafting fronds only play a transitory part.

If we were to look carefully at the underside of the fern fronds that create the carpet of vegetation in this forest, we would see dozens and dozens of tiny dots. At a glance, the fern might seem to be infected with something or under assault by tiny insects. But these are actually spore cases, natural pockets filled with innumerable fern spores. Each one is like a speck of dust, but a speck that holds the promise of another generation.

Ferns obviously play the numbers game when it comes to reproduction. When they release their spores, the tiny dots go everywhere. Some will find suitable ground and start to germinate, and eventually grow into a structure called a prothallium, which looks like a small green heart with a scraggly beard at the bottom. This is what sets the stage for the growth of the fern fronds, but it can't do so on its own. The prothallium has sex organs that require fertilization before anything else can grow. In fortunate circumstances, that fertilization

leads to the growth of a long, curled stem that eventually un-
furls into a fan of small leaves.

Such complications might seem to make ferns unlikely signs
of a recovering world. The way they reproduce is archaic and
complex. The plants don't get assistance from other organisms,
the way flowering plants or some conifers do. Ferns kept low to
the ground while entire towering forests grew up and thrived
around them. Ferns stayed a part of the understory, preferring
relatively cool and darker habitats created by the shade of the
great trees above. And that is precisely why they have thrived
and filled the landscape as other plants have struggled to take
root in the same terrain. Strangely, the aftermath of the impact only assisted ferns further. While other plants suffered,
ferns found themselves the recipients of good fortune. During
a time when sunlight was blocked for weeks, and reduced for
months, ferns that evolved to grow in low light were hardly
fussed. And now that the hordes of hungry, herbivorous dinosaurs were gone, the way ferns vigorously reproduced was not
a matter of maintenance but of spread. Wherever there was a
cool, moist space, ferns rapidly proliferated.

Ferns will continue to play the same role as a disaster taxon
for millions and millions of years to come, rapidly growing in
disturbed habitats stripped of their former diversity. A bur-
geoning of ferns is a sign that an ecosystem is recovering after
a major disaster. In the wake of intense volcanic eruptions or
other catastrophes that rapidly disturb established environments, ferns are among the first signs of regrowth. Ferns don't
need other organisms to spread their offspring, and, when
hungry herbivores are rare, the plants can grow without being

constantly cut back. Ferns have been primed by evolution to thrive when other species struggle. Life's diversity is a buffer against the inevitable crises that strike the planet, the connections between what was and what will be.

The forest that's beginning to take root in this fern-covered landscape, however, won't be like those of even a million years before. Rather than being merely a backdrop to the evolutionary dramas of animals, the Paleocene forest is being shaped by both the nature of the survivors and the absence of the creatures that once called these woodlands home.

During the end of the Cretaceous, the forests of Hell Creek were absolutely packed with a wealth of different plant species. Scraggly-limbed monkey puzzle trees, palms, dogwoods, cypress, and more all grew alongside each other. Flowering plants and conifers vied with each other for access to sunlight, while ferns and cycads continued to sprout up along the forest floor. But these forests were not thick. The Hell Creek forests did not have closed-in canopies that created a world of shadows and shade below. There were often sizable gaps between one tree and another, spread-out stands that created a sparse canopy and let plenty of light in for low-growing plants.

Dinosaurs kept the forest this way. Massive herbivores such as *Edmontosaurus* and *Triceratops* unintentionally shaped the forests through what they ate and where they walked. Often, young trees would be trampled or consumed before they had a chance to grow. And while the leaves and branches of adult trees were often beyond the reach of these dinosaurs, an animal like *Triceratops* could still push some of these plants over— creating more open space in the forests. Even the game trails dinosaurs created—opened by the herbivores and followed by the carnivores—maintained this equilibrium. Trampled and packed soil was harder for plants to take root in, dinosaurian pathways creating lines that wove through the forest and could even be seen from above through the trees. Large dinosaurs were unintentional ecosystem engineers, reshaping the Hell Creek world around them.

Now they're all gone. No dinosaur larger than the size of a raven survived the terrible pressures of the heat pulse and impact winter. A million years ago the average animal in this environment weighed about three tons. Now there are few creatures that weigh more than three pounds. There are no more giants that will cultivate the prehistoric garden.

Plants, of course, went through their own mass extinction. Conifers didn't fare nearly as well as the angiosperm upstarts. The long reign of conifers as the primary components of prehistoric forests has come to a close. Flowering plants are growing much faster than the conifers, filling up much of the available landscape with broad-leaved plants that are nestled close to each other. Without large dinosaurs to keep the forest open and bathed in sunlight, the relatively small number of flowering plants are rapidly sewing together a multistory canopy that shields much of the forest floor from the sun. There has never been a forest quite like this before.

The natural history of the surviving plants and the fallout from the more intense days of the early Paleocene helped to underwrite this riot of plant growth. Flowering plants could already grow quickly, a result of their competition with conifers during the Age of Dinosaurs. These plants were already poised to break new ground in the early Paleocene. But the products of the terrible post-impact events has helped to set angiosperms up for success, too. The ash and burnt vegetation from the global wildfires created a glut of phosphorus in the early Paleocene soils. This is a major boon to some new angiosperms on the scene, and as the surviving angiosperms have begun to grow and change, new forms are beginning to emerge. Plants related to beans and peas are starting to proliferate, and in the process these new growths are beginning to fix nitrogen into the earth—taking nitrogen from the air and bringing it into soil compounds that will further fertilize the forest's foundation. All of these changes favor the growth of other angiosperms over conifers and ferns—those early, questing shoots of flowering plants are slowly converting the forests to the soil conditions that best benefit themselves as other forms of plants take root in smaller pockets of this new forest. And this is only the beginning. A densely packed, shaded forest with multiple stories from root to canopy offers more habitat than any patch of Hell Creek woodland, and mammals are set to take advantage of these changes.

Little *Baioconodon*, padding through the Paleocene forest, plays a different role in reshaping the world after the great extinction. Her ancestors survived the post-impact disaster. In eons to come, her relatives will become apex predators— carnivores that could reach lion-like statures, some of the first great meat-eaters of the mammalian age. And part of the secret to her lineage's success crackles away between her fuzzy ears.

Within the broad sweep of mammalian evolution, the brain of *Baioconodon* isn't all that large. In fact, much of her skull seems devoted to eating. Large flanges of bone, the zygomatic arches, make room for reams of jaw muscles. Compared to later beasts that will carve out a similar living, such as the cats and dogs, her brain seems very small relative to her body size, a ratio sometimes taken as an indicator of relatively low intelligence. Some of the largest dinosaurs, for example, had brains that you could fit in the palm of your hand, with much of the processing capacity seemingly devoted to smell and sight rather than cognition. Early mammals were similar, their brains best suited to finding the next meal and the requirements of reproduction rather than complex behavior.

But the asteroid strike nevertheless changed the fortunes of mammals, not just in terms of body size but also their brains. *Baioconodon* and her neighbors are already as large as the last mammals of the Cretaceous—a fairly speedy recovery given that only the smallest mammals were able to survive the post-impact trauma. And their brains are not only becoming larger, but larger in ways that allow for new behaviors to evolve. No longer constrained by life in the shadow of looming reptiles, mammals are plowing new niches in the Paleocene world. As they do so, variations in what each mammal can do—from the food they eat to how they interact with others of their kind—get nudged this way or that by natural selection, opening a variety of new behaviors. The open landscape doesn't result in a homogenous group of beasts, each doing more or less the same thing. As the Paleocene mammals continue to evolve through the generations, novelty spawns more novelty as niches become subdivided and some species become more specialized. An omnivorous ancestor might give rise to carnivorous or herbivorous descendants, and the arrival of a new herbivore or carnivore on the scene means that the constant churn of evolution will spin off even more variations.

It's not just the absence of so many great and terrible dinosaurs that's made these changes possible. It's also the mass extinction of Mesozoic mammals at the end of the Cretaceous.

The ubiquitous clawhold of dinosaurs on the world's terrestrial ecosystems undoubtedly affected mammalian evolution just as it altered the world's forests. But we should be careful not to give too much power to the terrible lizards. They were not the only creatures around, and certainly not the only creatures that wielded ecological influence. Even though the reign of the dinosaurs put some sharp constraints on how large mammals could become, it was competition between mammals that limited the number of different forms and niches Mesozoic mammals evolved into.

During much of the Mesozoic, the most numerous beasts belonged to groups categorized as *mammaliaformes*, or mammal-shaped. They were not just like today's marsupials or placental mammals. Mammaliaformes were more archaic, and some still kept the ancestral trait of laying eggs instead of giving birth to live young. It really wasn't until the Cretaceous that therians—the group containing marsupial and placental mammals—evolved and started to make their mark on the landscape. There was just one problem: many niches available for the new mammals were already occupied by the more ancient mammaliaformes. This is the same issue that early dinosaurs faced back in the Triassic. The first dinosaurs coexisted with their more archaic precursors, and it wasn't until those more ancient forms of reptiles went extinct that dinosaurs really began to flourish.

The mammals that became the Mesozoic equivalents of

beavers, flying squirrels, aardvarks, otters, and more were primarily mammaliaformes. The earliest therian mammals, however, were little insect-eaters, some of which were so tiny that they weighed less than 100 grams. Therian mammals, in other words, evolved in a world where many of the ecological roles possible for mammals were already occupied by more primitive forms. The early mammals diversified, but they didn't become disparate, or, the growing number of species were more or less the same to each other. There would not be therian equivalents to mammals eating fish, flying between trees, or digging through ant nests for millions of years yet. Just as flowering plants could only become more firmly rooted when the mass extinction cleared some of the competing conifers out of the way, the ancestors of cats and dogs, bats and primates, elephants and whales, and so many other forms of mammal could not even begin to establish themselves until the beasts of the older establishment were ushered away. *Baioconodon* is part of the new guard, a mammal with molars that both crush and shear, capable of giving birth to live young gestated inside. It's this type of anatomical foundation that will lead to a new flowering of mammalian evolution, setting the stage for a very different array of creatures than might have been if the more primitive mammaliaformes had held sway.

Above the head of *Baioconodon,* hidden within the branches of a broad-leaved *Meliosma* tree, a small bird makes a *chp chp chp* sound. It's a small avian with a short, pointed beak and a short splash of feathers rising up between its eyes. It's not the most colorful creature, mostly a buffed brown with lighter feathers underneath, but the little singer is a sign that even the dinosaurs are beginning to recover. The little bird is one of the earliest mousebirds, a family of fluttering dinosaurs that will stick around for tens of millions of years to come.

The birds that survived the pressures of the mass extinction were not archaic or primitive. In fact, many of them were the earliest members of lineages that would continue on through tens of millions of years. Just like mammals, birds were diversifying and splitting off into new lineages even during the time non-avian dinosaurs filled the land and pterosaurs competed with birds in the air. The earliest relatives of parrots, waterfowl, and other birds had already taken up their perches before the asteroid struck, and, being beaked birds capable of eating seeds and nuts, many have been able to survive into the days of the great Paleocene recovery. They are the dinosaurian legacy. Just like their imposing relatives, the surviving birds grow fast and produce multiple offspring each year. It won't be long before six-foot-tall flightless birds start to stalk the forests, happy to feast on whatever still-small mammals might run across their paths.

But the mousebird of this particular Paleocene dawn is no such terror. The bird is part of a small-scale profusion that again fills the forest with song. No non-avian dinosaur could make a sound quite like it. There are two reasons for that. The first is a specialized organ called a syrinx. The organ looks like a corrugated tube that branches off into two parts at the base of the larynx in many of the surviving birds. The structure evolved millions of years before among the beaked birds of the Cretaceous; it allowed Mesozoic birds to squawk, chirp, and otherwise vocalize in ways far different from the deep, guttural noises of the non-avian dinosaurs. And just like the mammals, the Paleocene birds are starting to evolve new behaviors and ways of interacting.

The brains of the Paleocene birds are starting to become bigger, with more of their cerebral wiring devoted to cognition and behavior. The birds are not running on autopilot, but considering and interacting with their world in ways the non-avian dinosaurs never could. Contrary to the history of their now-dead relatives, though, birds are getting this benefit because they're getting smaller.

Flight is a costly way of getting around. The amount of energy required to take to the air and stay there is staggering, especially as birds become larger and larger. A big bird has to take advantage of thermals and the ability to glide, just as big pterosaurs did during the Cretaceous, in order to make a living. Not only that, but many of the niches opening up in the Paleocene world are within hot, dense forests. Each stand of trees, even each tree itself, offers a variety of different habitats. There are birds that spend a great deal of time picking morsels out of the soil, birds that make their homes in tree trunks, birds that hop around the upper branches, and more. This is a complex world that favors small species, and birds are already predisposed to shrinking over evolutionary time. In fact, the very first birds were the result of miniaturization—their dinosaurian ancestors were much larger, about turkey size, and evolutionary pressures related to flight favored smaller fliers, leading to the evolution of birds that were even smaller still. But the brains of birds didn't get smaller as their bodies did. Their brains remained the same size, opening up the possibility that parts of that valuable processing space could be freed up to handle songs, mating dances, and a much busier social life than non-avian dinosaurs ever had. And that wasn't all. While the carnivorous ancestors of birds relied a great deal on scent, most beaked birds lost their ancestral keen sense of smell over evolutionary time. Sight became more important, even seeing in the ultraviolet parts of the spectrum. The way *Baioconodon* might see a mousebird, and the way mousebirds see each other—little UV freckles dotting their chests—are very different.

Against the background of all this evolutionary change and potential, though, some creatures will ply a more conservative course. Not far from the little chittering mousebird, on the sandy bank of a quiet stream shaded by overhanging tree branches, there rests a little crocodylian. Its snub-nosed look makes it seem like a pugnacious caiman, which is not too far off from the reptile's place in the evolutionary tree. A dark brown above, and a lighter cream below, the ectotherm simply lies on the bank with his jaws agape, waiting until something catches his attention—or blunders into the sensitive flesh of his toothy mouth.

The recumbent reptile isn't quite like any that came before or those that will come after. He's part of a particular Paleocene species. Still, his shape is awfully familiar. His skull is low, with eyes and nostrils given a little extra elevation so that he can keep them above the water when otherwise submerged. His feet are webbed, and under many of his pointed back scales are tough osteoderms that act as a natural body armor against the bites of other crocodylians. He wouldn't look out of place in a Jurassic swamp, and yet here he is in the Paleocene.

It would be unfair to call such a creature a living fossil, however. Crocodiles have a long and varied history, even through the Mesozoic. The earliest crocodiles, technically known as pseudosuchians, were terrestrial creatures. They ran around on land and competed with the earliest dinosaurs for ecological prominence. After the end-Triassic extinction decimated their family, some of the survivors became small and mammal-like—even evolving dental tool kits of varied teeth to tackle particular prey—while others became adapted to life in the sea and still more took up the life of semi-aquatic ambush hunters. The dinosaurs were relentless, rarely giving the pseudosuchians a chance to gain much purchase in the Jurassic world, but the identity of what a crocodile *is* was very different between 235 and 66 million years ago. Even by the time the asteroid struck, there were odd crocs living in some pockets of the world. In Late Cretaceous South America, for example, there stalked terrestrial crocodiles called sebechids with curved, knifelike teeth. These carnivores filled the same niche as medium-sized predatory dinosaurs elsewhere, only to perish with most theropods in the aftermath of the impact.

The only surviving crocodiles were those tied to water. All the terrestrial species vanished, but the sneaky, semi-aquatic species fared far better. Just like the dinosaurs, the abundant and diverse forms of their family were winnowed down to a much narrower selection. And this is the way things will stay. There will be the occasional terrestrial croc in the millions of years to come, some of which will even have hooflike feet to better run after the mammalian prey of the semitropical forests. But for the most part, the waterlogged species are the ones that will carry the crocodylian banner forward through time.

Time and again, the surviving crocodile lineages will evolve similar appearances. Some will have triangular snouts of crocodiles, others the rounded muzzles of alligators, and yet more will have long, narrow jaws best suited to catching small fish. These skull shapes all relate to diet, and varied crocodile populations will be nudged between these shapes over and over through time—rapid evolution to arrive at a small set of familiar shapes. That's why crocodylians, like this Paleocene caiman, seem so ancient. Different creatures have been lurking at the water's edge, ready to snap up the unwary, since the very first amphibious vertebrates dragged themselves onto ancient mudbanks. After them came alligator-like amphibians that shared many of the same adaptations. After those giant salamanders vanished, reptiles—vaguely crocodile-like animals called phytosaurs—filled that ecological space. When the phytosaurs disappeared, crocodiles snuck into the shallows and there they have been waiting ever since. There's hardly any reason for them to undergo a major evolutionary jump. Animals everywhere need water. Crocodylians can easily hide in even a few inches of pond or stream, and their slow metabolisms allow them to survive a long time on relatively infrequent meals. Through evolutionary happenstance, they sidestepped the reptilian rat race of endothermy. If you can survive on little, in an environment buffered from catastrophe, there is little reason for change beyond adaptive tweaks or whatever flashy features allow a crocodile to leave behind a larger genetic legacy. Crocodylians aren't waiting for the Age of Reptiles to come again. They happened upon something that works, and there's hardly a reason to make aspirations beyond the marshes.

A buzzing dragonfly, searching for invertebrate prey, flits along the same stream bank where the croc rests. Its movements are erratic, at least to vertebrate eyes, and the lounging reptile doesn't seem to take any notice. But a sudden breeze pushes the hovering red insect just far enough to the left. *Crunch.* The crocodile smacks its jaws, in a moment dislodging the insect from the tooth it became impaled on, swallowing the crunchy prize before settling his chin down on the sand. Such reflexes can't be helped. Despite his heavily armored appearance, he's a sensitive creature. His jaws are positively dotted with almost invisible bumps called integumentary sensory organs. Sometimes he merely waits in the water, jaws open, and a fish tickles the little sensory dots just so; the great muscles of the reptile's skull pull to close those jaws in the blink of an eye. They make hunting a snap.

The dots aren't just scattered over the reptile's jaws. There are also a few here and there along the sides of his body. On a day like this, they sense the water—pressure changes that might indicate movement. In mating season, however, touch becomes very important to his species. Mating follows courtship, and courtship involves two of the crocs sliding alongside each other's bodies, pressing against each other. It's a reptilian romance that's been playing out every season for millions of years, an unexpected ritual made possible by adaptations for a life semisubmerged.

The forest begins to grow quieter as the sun rises higher. The nocturnal species are now asleep. Many of the diurnal species seek shade from the hottest parts of the day. It's warm again, warmer than it was during the last days of Hell Creek. This isn't the deadly, singeing heat of the post-impact days, but the beginnings of a new climate shift that will only ramp up further in the next 9 million years. The sun-reflecting compounds of the impact winter have largely left the atmosphere now, and the lingering carbon dioxide, methane, and other greenhouse gases—some recently spewed out by the continued belching of the Deccan Traps—have begun to trap even more heat. By the end of the Paleocene, there won't be any more ice caps on the planet and crocodylians will slide through swamps filled with *Metasequoia* just a stone's throw from the North Pole.

There's no single point where life can be considered fully recovered. There will never be creatures as large and prolific as the non-avian dinosaurs ever again. Mammals cannot grow that large, not only because of physical constraints but because most left egg-laying behind. A smaller range of body sizes can translate into fewer ecological roles to play, with fewer hangers-on. The world will never be the same. Even so, life is now pioneering into new crevices, new cracks; the diversity of species on the planet is—on the whole—increasing rather than decreasing. Growth is a sign of life, after all, and Earth once again feels very much alive.

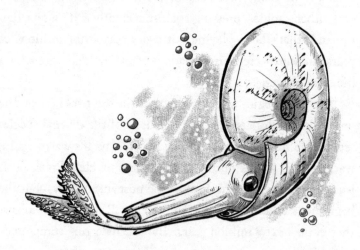

Elsewhere

SOMEWHERE IN THE MIDDLE OF
THE ANCIENT ATLANTIC OCEAN

The speck is one of an infinity of tiny dots floating through the blue. The creature is another bit of plankton in an ocean that is slowly recovering from the injury of mass extinction. But this pinpoint carries a tragic uniqueness among all the other tiny plants and animals floating along the ceiling of the ocean realm. It's one of the last ammonites.

Up close, the oceanic dot is a miniature cephalopod so intricate that even the finest artists would be impressed with the delicate nature of something so tiny but so brimming with potential. It's a tiny *Pachydiscus,* once one of the most prolific ammonite species in the entire world.

Ammonites are relatives of squid, similar in appearance to the pearly nautilus while belonging to an entirely different family, the ammonoids, an offshoot arm of the cephalopod family that appeared around the time that the earliest dinosaurs were trotting around on land. They were an evolutionary hit. Bobbing coil-shells could be found in every ocean, ironically being invertebrates that supported the backbone of marine ecosystems. The smallest could rest comfortably on your pinkie fingernail. The largest were larger than a compact car. Some had simple, smooth shells. Others carried ridges. Some had fractal suture patterns dictated by the mathematics of biology. Still more had shells that uncoiled and took on bizarre trumpet shapes. They were so prolific and adaptively energetic that paleontologists would one day use them as biomarkers to compare rock layers around the world: particular species occur for narrow strips of time but are found in enough places that telling the age of a particular layer isn't always a matter of geology but of coordinating mollusks from around the planet. Altogether the ammonites survived at least one mass extinction and several other minor disturbances, their place in the seas seemingly assured.

But now there are few. The tiny ammonite floats in a sea that is as vast and mysterious as space is to us. It looks like an extremely small version of the adult. Adult ammonites produced vast amounts of larvae that became part of the ocean's plankton. These babies looked like itty-bitty adults, their suckered arms dangling from simple curled shells. In time, a lucky ammonite would build more shell as it aged. It's the cephalopod version of buying a simple one-room shack and gradually adding more and bigger rooms over time, the soft body of the ammonite living in the newest and largest chamber as it grows in a curved pattern around the old. The empty valves become critical for staying afloat or sinking as the situation dictates, gas held in those chambers or expelled from them allowing ammonites a clunky form of mobility. The sturdiness of their shells was enough to protect them from any unexpected hard knocks.

This larva is not so lucky. There's something wrong with its shell, which seems to bob in an awkward, jerky pattern with the motion of the water. Up close, the small, older chambers of the shell have holes in them as if someone scraped away at them with something hard or otherwise sanded them down. This isn't a bite or some disease. There's no sign of infection or damaged soft tissue. The shell's been eaten away; eaten away by the ocean itself.

Catastrophic as the immediate effects of the asteroid were, the event started a slow buildup of changes that were equally deadly in other ways. The cataclysm was not just one of fire. Vast quantities of carbon were released into the atmosphere and subsequently absorbed by the ocean. A side effect of this mixing is that the ocean's pH levels dropped. The waters didn't become a roiling sulfuric stew. The change may be all but invisible to the sea's inhabitants. But the effects of the ocean acidification are clear. Shelled organisms—ammonites, giant clams called rudists, and more—struggle to build shells in an environment that's actively working to break down those same organic materials. Baby ammonites cannot make their shells properly. Those that don't perish in their larval stage become easy prey and inefficient predators in their own right, their shells pocked with holes.

The ammonites aren't consciously aware of the change. They swim, feed, and reproduce as they always have. Even at the time of impact, their kind was spread around the world in every ocean. By all logic they should have been buffered against extinction. Having a wide range offers an advantage that perhaps some population will find a safe haven somewhere, a refugium where they might persist even as their relatives falter. But there is no such protection for the ammonites. With each generation, their numbers now dwindle.

10

One Million Years
After Impact

Nights have never been the same since the reign of the non-avian dinosaurs ended.

Feathery avians can still be heard in the darkened forest. Descending notes—*woo-oo*—set the tone of the dark hours as some of the nocturnal birds sing to each other. Call and response, call and response, a little ping that lets each of the unseen avians know where each is and where invisible territorial lines lie.

The intermittent noise is nothing more than a distraction to the *Eoconodon* snuffling along the forest floor. The birds perch in the branches overhead, but the squabs are out of reach for the German shepherd–sized omnivore, who lacks claws suitable for scaling the rough tree bark. Nose to the ground, sniffing with each short step, the mammal seeks out a late-night snack elsewhere.

Compared with the carnivores of the Late Cretaceous, *Eoconodon* would barely make a mouthful for a *Tyrannosaurus*. Even the largest of these fuzzy quadrupeds weighs only about eighty pounds soaking wet—a common occurrence in the sweltering forest. But the days of the tyrannosaurs are long gone, to say nothing of the hadrosaurs, horned dinosaurs, ankylosaurs, raptors, and all the other terrible lizards. All that remains of dinosaurian existence are the chatty birds. *Eoconodon* is now the largest carnivorous mammal. The only meat-eaters that outsize it are the snub-nosed crocodylians that bask along the riverbanks and sit motionless in the tea-colored water, endlessly patient for the right moment to clamp their jaws shut on the unwary.

Mammals like *Eoconodon* had not existed a million years earlier. Nor would they be around for long after the time of this stubby hunter making its way through the moonlit undergrowth. *Eoconodon* is a mesonychid, a member of a group of mammals that would come to be known as "wolves with hooves." These sharp-toothed mammals don't have the blunted cleats suited to endurance running that true wolves would later evolve, nor do they have curved claws to snatch at prey like cats. Instead the feet of *Eoconodon* are tipped with blunted, keratin-covered nubs that provide grip on the sodden forest floor.

But the bite of *Eoconodon* is more powerful than its cute, roly-poly exterior might suggest. The mammal's jaws are long and doglike, large, conical canines jutting down from the upper and lower jaws. This isn't a new mammalian invention—mammals had been nipping with differentiated teeth since the time of their origin over 150 million years earlier—but *Eoconodon* is one of the first to take the dental armaments to a new extreme. Without hulking dinosaurs ruling the roost and keeping mammals small, little omnivores are able to get larger and start dining on a great variety of morsels, including other mammals. The predator niche is wide open, and *Eoconodon* is one of the first to fill it. And in doing so, the lineage of larger and larger mammals that produced *Eoconodon* put jaw strength before brain power.

From the outside, *Eoconodon* looks something like a cross between a dog and a raccoon. From a narrow snout, the mammal's skull widens toward the back of the jaw. But if we could look inside this beast's head, we'd find more muscle than brain matter. The brain of *Eoconodon*—like those of all Paleocene mammals—is relatively small. The body's neural control center is cradled in a small island of bone behind the eyes. In other words, a modern house cat would be able to handily outsmart *Eoconodon*. What matters more in this mammal's time and place, however, is crunching power. The cheekbones of *Eoconodon* swing out wide from the jaws, and that small braincase creates two big holes for the jaw-closing muscles to thread through. These are the apertures for the masseter muscles, the flexing and bulging bundles that allow prey to be dispatched and bones to be crunched against the molars. Even though the teeth of *Eoconodon* lack some of the specialties of later carnivores—such as the shearing bite later pioneered by cats and dogs—the mammal is certainly capable of crunching most of its neighbors.

But even the great meat-eaters of the Mesozoic took millions upon millions of years to evolve. The low-slung, shuffling form of *Eoconodon* makes sense given that just a million years before, her ancestors were scurrying around the Cretaceous forests in search of insects and lizards, trying to avoid the curved claws of raptors and the terrible jaws of little tyrannosaurs. This mammal is both a complete, wonderful creature and one that, unknowingly, is part of an evolutionary continuum. Some of her traits are holdovers. In addition to teeth that are capable of chewing meat, even if they were not specifically adapted for it, the ears of *Eoconodon* are round and small. This mammal doesn't have the large, swiveling ears of later dogs and cats that will help those carnivores pinpoint prey. Instead, much like its ancestors, *Eoconodon* is more scent-focused. The olfactory bulb at the front of the mammal's brain is larger than the rest, a processing center for all the invisible aromas that form trails throughout the mammal's humid jungle home. All *Eoconodon* has to do is follow her nose.

On this night, however, suitable prey is hard to find. *Eoconodon* walks to the edge of the forest bank, turning to follow the quiet forest stream. The silvery moonlight glints off the wet back of a soft-shell turtle. That'd be easy pickings, but in the dark *Eoconodon* might blunder into a hidden crocodile. Even a small one can leave a nasty and debilitating bite. The mammal continues on, snuffling air through her wet nose to pick up the pungent scent of another mammal or perhaps a ripe, soft carcass decaying on the forest floor.

As the hungry mammal shuffles along, her search is becoming more and more frustrating. *Eoconodon* can sense where prey *was* but each trail seems to turn up empty. The promising scent of *Carsioptychus*—a similarly sized mammal that prefers to munch leaves over flesh—leads to nothing but a small pile of chawed twigs and half-eaten leaves. The scent of a rotting river turtle only turns up the footprints of another *Eoconodon*. The hungry wanderer squats over the spot, lifts her tail, and rubs against the sandy earth. She'd have better luck if other *Eoconodon* stayed away.

A warm nighttime breeze flutters for a moment, buffeting leaves and flexing branches. Something just above her head catches her eye. Long thin pods hang from a low branch, little botanical parentheses attached to thin branches covered in frond-like leaves. These smell different somehow. While *Eoconodon* is not unwilling to supplement her diet with leaves, most plants are just not as flavorful as a mouthful of *Crasioptychus* flesh. But there is something about these pods that inspires her to shuffle back a step, push back with her forefeet to balance on her hind legs and tail, and bite into one.

The pod itself is unsatisfying. It's tough and chewy, fibrous and not very palatable. But inside, there's a treat. Firm, fleshy little dots that seem somehow reminiscent of meat. She rears back to grab another, this time holding down the pod while messily shredding through the exterior to get to the tastier morsels inside. Just as the mesonychids are a novel form of life on Earth, so is this meal. It's one of the earliest legumes—an ancient bean plant.

Plants were as affected by the end-Cretaceous disaster as other forms of life. Plants, stalwart and permanent as they seemed, struggled to survive through the fire and the long dark that followed. The catastrophe was as much a reset for forests as it was for mammals, and now angiosperms have the advantage. For millions of years, flowering plants pushed against the dominance of conifers and their relatives. Every patch of open ground in the Mesozoic forest was contested, everything from sunlight to soil nutrients affecting what species could take hold and have a chance to proliferate. Against this background, with flowering plants ascendant, the stage is set for the origins of legumes.

Not all plants are equally nutritious. Some are tough and fibrous and offer little in the way of nourishment. Others are sweet and sugary, a treat but not something that can sustain the likes of *Eoconodon* full-time. And plants often evolve defenses—burning sap, stinging hairs, and silica-rich leaves that can wear down teeth fast—to keep themselves from being cut down by hungry creatures. Some plants nourish and others irritate. Mammals have to be choosy lest they wind up with swollen stomachs of irritating, indigestible leaves.

Eoconodon has chosen wisely. The beans inside the tough outer pod have nine times the amount of protein of the little wisps of grass growing along the stream bank. Not that the mammal knows this or anything about nutritional value. The beans just taste good, their novelty being an excellent spice. But the fact that this plant is growing here, now, is a good sign for the future of mammals, a whisper of what's to come.

Living large doesn't come free. Constraints, trade-offs, and the question of acquiring adequate fuel all have their roles to play. During the heyday of the dinosaurs, truly large and stupendous sizes relied on a few factors. All dinosaurs laid eggs, starting off life small. By outsourcing gestation and avoiding carrying around increasingly large infants inside, dinosaurs decoupled limits on size from reproduction. On top of that, many of the largest dinosaurs had light bones—systems of air sacs extending outward like roots from their respiratory system, keeping bones light without sacrificing strength. And the largest species of all—herbivores stretching more than one hundred feet long and weighing more than seventy tons—evolved long necks tipped with heads that acted as garden shears, clipping vegetation that was swallowed whole and processed in the stomach and hindgut. All that anatomical equipment was important in the age before angiosperms became widespread. Flowering plants didn't evolve until well into dinosaur history—during much of the Mesozoic, most of the available fodder were ferns, cycads, and conifers, not exactly known for being packed with caloric energy. Giant herbivores had to eat and eat and eat to get their sustenance in bulk, growing sufficient flesh for larger and larger carnivores to consume.

Eoconodon, its relatives, and its successors cannot follow the same pathway. Evolution will have to play out in a different manner. While a few mammals—the monotremes—still lay eggs, they are evolutionary holdovers from tens of millions of years before. The once-dominant marsupial mammals, too, were hit hard by the mass extinction, there being fewer species carrying tiny joeys in their pouches. This is a time for the placentals, mammals like *Eoconodon* that carry their young inside their bodies to term. While beasts would not reach their largest terrestrial sizes for another 40 million years, carrying babies inside the body puts a lower ceiling on how large mammals would eventually become. To get bigger, placental mammals have to carry larger and larger offspring, which requires longer gestation, which involves greater chances of something going wrong in the process. The babies are kept snug and safe, but bigger bodies requires bigger babies.

And those bigger bodies run hot. *Eoconodon* and its mammalian neighbors are endotherms. They maintain their body heat at a high, constant temperature, generated from the inside. The mammals do not need to lie out in the sun to warm up each morning like the crocodiles or the little crunchy lizards *Eoconodon* sometimes nabs as the reptiles go about their sluggish morning routine. The mammals are always quick, always ready, but maintaining such an active state requires more fuel. Simply scarfing down any plants in the vicinity won't do. Mammals need energy-rich foods that provide more nutrition for each chew, and they need to find them every day. Gorging might hold over the likes of *Eoconodon* for a day or two, but the fuel goes fast. Without sufficient salad in the surrounding environment, the possibilities of growing larger would stay closed.

That is why the appearance of simple, unassuming beans is so important. Beans provide energy to move, energy to reproduce, energy to grow. All animal life on Earth is underwritten by the generosity of plants.

Even the insects are enjoying the relative bounty of the Paleocene forest. Many of the leaves in this patch of warm woodland bear circuitous little squiggles—the hallmark of leaf miners. Insects that started off life this way—multiple species of flies and beetles—were hit hard by the devastation of Cretaceous forests and the long impact winter. Even if there was no broadscale extinction of arthropods, species that were tied to specific plants perished. But insect evolution has a habit of re-peating itself. Even though many of the insect species that re-lied on plants for their larvae vanished, new ones have sprung up in their place.

One such larva is at work in the leaf of a plane tree, grad-ually winding its way through the broad, veiny foliole as the young insect feeds. If this insect is lucky enough to survive, it'll eventually cocoon itself in silk, pupate, and emerge as a moth—a fluttering insect full of fatty compounds that the forest's small mammals would likely consider a satisfying, if somewhat dusty, snack. For now, though, the little caterpillar is merely chewing away, creating thick, wobbling lines along the margins of a thick leaf vein, kept snug as a bug inside the leaf's protective layers. It's not a perfect defense. Parasitoid wasps can still come along and, if they find the larva, inject their eggs into the caterpillar's body. The wasp larvae will grow, kill their host, and eventually burst out of the leaf mine before it gets a chance to leave its snug chamber. Yet, gruesome as this may be, it's yet another sign that the world is rapidly recovering from the destruction of a million years before. The moth larva is not a survivor from the Creta-ceous, but a new species that has evolved in the Paleocene. The wasp, too, can only survive as long as there are leaf miner larva to parasitize. Each component of the ecosystem—the plane tree, the leaf miner, the wasp—is tied to others, recovering the complex layers of dependence that had been shattered when the asteroid struck.

Naturally, *Eoconodon* takes no notice of the little leaf miner above her head. She's focused on filling her own belly, even if the plants are not quite as satisfying as fresh flesh. Before long, *Eoconodon* has her fill. She gives a last lick to one of the now-empty bean pods lying between her broad paws, hoping for one last bit of flavor. Not quite as satisfying as fresh meat, no, but it will do, and at least the bean pods can't run. With a great stretch, she leans back and opens her jaws wide, her flexible nose tilting up to further expose her large, conical canines. Somewhere nearby, out of reach of those teeth, a tiny mammal shouts a rolling chitter—an alarm call.

Eoconodon looks up. There, perched on a thin branch bobbing with the weight of the little mammal's irritation, is a small beast that looks like a cross between a shrew and a squirrel—*Purgatorius*. The small mammal chitters again, longer this time, the call repeated by another *Purgatorius* unseen in the nighttime forest. *Eoconodon* snorts and turns away from the irate creature. The fuzzy morsel is too far out of reach to nab, which makes its persistent chattering and bottlebrush tail seem like an overreaction to the carnivore on the forest floor.

With another snort and a shake of her broad fuzzy head, *Eoconodon* moves off. The night is starting to pick up hues of gray and blue. Morning will come soon, a good time to get some shut-eye. But the *Purgatorius* is not thinking of sleep. She has other matters to attend to. After another few swishes of her tail, she scampers back to the trunk of the tree and scurries inside. Good, they're still there: two little dots covered in thin fur—her babies. She sniffs them, licks them, and lies with her warm, soft belly toward her curled-up babies, still something new in this world—primates.

The very first primates did not wait for the likes of *Tyrannosaurus* and *Torosaurus* to totter off the evolutionary stage. Around 67 million years ago, as the Age of Dinosaurs seemed ensured for epochs to come, a new sort of mammal started running through the trees. They looked something like tree shrews—long bushy tails, triangular muzzles fitted with teeth suited to catching insects and munching fruits, and clawed, grasping hands and feet capable of swiveling to latch into the tree bark. These were the first *Purgatorius*, small beasts of the canopy and the first of their primate kind to exist on Earth.

They could have easily perished in the cataclysm caused by the asteroid. The very nature of *Purgatorius* was tied to the trees. The little primates needed tree holes to sleep in, fruits to eat, and the insects that hid themselves among the leaves and branches. The homes of *Purgatorius* were little more than kindling as the atmosphere turned to an oven and wildfire took much of what was left. Many of their species were lost. But enough survived, just enough to keep reproducing, to keep searching out the next meal in the wreck of the old world. Where the mighty *Alamosaurus* and the cunning *Atrociraptor* perished, *Purgatorius* somehow survived.

The little primate is no monkey. *Purgatorius* does not have an opposable thumb. The little primate is not even particularly brilliant compared with the other mammals of its time. To look at *Purgatorius*, without knowing what would follow, we might think that she is just a twitchy little insectivore like the many before her and the many that will follow. But we have the privilege of knowing better. Safe and sound with her little ones, she is one of the oldest members of the plesiadapiforms—squirrel-like primates that began to fill the treetops as the likes of *Eoconodon* and *Carsioptychus* made the forest floor their home.

There is nothing inherent about *Purgatorius* that explains why her story is one that will become important in the fullness of time. Small mammals had been living in the trees and nabbing insects for millions upon millions of years already. With dinosaurs thick on the ground, the trees were a relative refuge—or at least a place full of hidey-holes and ways to escape that even the feathered dinosaurs were not as adept at navigating. *Purgatorius* was just another survivor from the Age of Dinosaurs that managed to keep its treetop hold in the next great chapter of life's story. She is a mammal in the right place, at the right time, facing little competition for all the crunchy, sweet goodies in the trees. And she can eat them all. She is not specialized on one type of beetle or fruit. The pace of her metabolism runs fast. She has to feed more often than *Eoconodon* does, especially with a pair of babies that are almost constantly hungry for milk. She takes what she can, turning her world into the energy she needs.

But this Paleocene world is hardly a paradise. There are still threats here. Even though she need not have worried about *Eoconodon,* little *Purgatorius* has been awakened by big shining eyes and a large toothless beak peeking into her den before. The lizards and snakes that go unnoticed by the larger mammals also sometimes blunder onto her doorstep. A few alarm calls and quick nips to their scaly hides soon send them on their way. The cost of survival is vigilance. She could pass that much on to her offspring. They will not stay with her very long. When they are sure-footed enough to catch their own food, she will nip and chatter and drive them away from the nest, just as she had done with her previous litters. This isn't from hard feelings, but simply what evolution has driven her to do—natural selection favors a path where these primates will venture far from home, breed with offspring from other families, and create new mixtures of genes that will be further modified through the generations. But for now, she could be their world—the warmth and sustenance the babies could not find anywhere else.

At this particular moment, however, the rise of the mammals is anything but assured. The world is recovering, rapidly becoming a lush and complex planet of blue and green once again, but just because the non-avian dinosaurs and pterosaurs were eradicated does not necessarily mean that mammals will have an unquestioned dominion. Many reptiles—from the soft-shell turtles that stick their pointed noses above the stream waters to the beaky birds—still remain, and they could very well begin a second Age of Reptiles. Mass extinctions do not dictate winners. There is no such thing as victory in this forest. When a global disaster ends one evolutionary dance, shifting the tempo, another begins, with no certainty as to who will lead.

But a great resurgence of reptiles is unlikely. The effects of the impact have left behind reptiles very different from those that were present in the Triassic, creatures that in many ways are more specialized and with particular constraints that will not allow them to easily evolve again into forms like *Tyrannosaurus*, *Edmontosaurus*, or *Alamosaurus*. Even though birds are the dinosaurs that survived, they will not be able to recapture the staggering array of forms that their relatives evolved into during the Mesozoic.

All the birds in this forest, all the species that survived, can fly. They are the last survivors of an entire array of flapping and fluttering dinosaurs, notable for their toothless beaks, lack of a tail, and fingers rearranged into wings incapable of supporting grasping, catching claws. In order for the chattering, swooping birds in this forest to evolve into something like their relatives, they would have to be shunted through some dramatic evolutionary changes. For one thing, some birds would have to become more terrestrial, more suited to efficiently strutting around on the ground than flying through the air. That change will indeed occur, and will be one step toward something a little more like a tyrannosaur. Likewise, these terrestrial birds will have to get bigger in order to compete against the mammals of their time and protect their nests from thieves—another change that will happen here and there. But that's where the circumstances of anatomy and ecology begin to box birds in. Birds lost their long, bony tails long ago. That's a major barrier to birds becoming like many of the quadrupedal dinosaurs of the Mesozoic, given how important a tail is as a counterbalance and appendage needed for steering. Nor can birds evolve to have teeth or claws. The genes for these weapons and tools are still in their bodies, but they have been shut off by mutations and slowly degraded over time. There is no evolutionary switch that can just turn them back on. Some bird beaks will evolve to be toothlike, and a rare few birds will retain claws for climbing tree trunks, but the snatching hands of the raptors and the terrible mouths of the tyrannosaurs will not evolve again. When birds grow large—especially in places where there are few terrestrial mammals that might eat them—they will do so in their own way and not as a return to the nature of the Mesozoic.

And what of the rest of the reptiles, the champsosaurs, lizards, snakes, turtles, and crocodiles? Some, in time, will indeed grow to stupendous sizes. About 5 million years from our present moment, in other Paleocene forests, there will be snakes and crocodylians that exceed thirty feet in length—swamp-bound creatures of a sweltering global hothouse. But that by itself is part of their secret. The surviving reptiles are species that were able to burrow or were semi-aquatic, and largely ectothermic—their body temperature determined by the surrounding environment. In order to become highly active, warm-blooded creatures, the survivors would have to go through some dramatic changes. For the dinosaurs, for example, the fact that their ancestors became small, ate insects, and were covered in feathers was tied to their hot-running metabolism, offering more evolutionary options once their ecological competitors died off. But now, even a million years into the Paleocene, any reptile undergoing such changes would face threats from the warm-blooded mammals that are already proliferating through the landscape. Many reptiles survived because of their slow, semi-aquatic lifestyle, and to compete with mammals they'd have to go through a spate of changes that might be impossible when the forests are already filling up with beasts. Some crocodylians will indeed go through such changes—once again becoming terrestrial predators that stride about on column-like legs, their jaws gleaming with knifelike teeth—but, in the long run, they will be more like passing fads against the background of mammalian proliferation. For both birds and semi-aquatic reptiles, there are more evolutionary steps involved to become ecological centerpieces like the non-avian dinosaurs were, and mammals got there faster—just as the ancestors of dinosaurs did when protomammals were devastated.

But it would be a mistake to simply focus on the legacies of the surviving vertebrates alone. These creatures only make sense in the context of ecology. And it's the spread of forests—especially the flowering plants—that has delivered mammals opportunities unlike any they have encountered before. The growth of the Cenozoic world and all that will come to thrive within it is built upon the foundation of the forest.

The story began long before the K-Pg impact, during the days of the Early Cretaceous when angiosperms were new. Flowering plants did not exist for almost all of mammalian history. And even after they originated in the Early Cretaceous, they were not all that common. The mass extinction is what allowed flowering plants to really take off. And with flowering plants came insects.

Mammals need to be thankful to insects for at least two reasons. The first is that more insects mean more sources of crunchy food. As Paleocene insects have undergone their own evolutionary radiation, replacing lost species and adjusting to the shake-ups in the forest, they are spinning off new species and beginning to fill the world again. They have their own ecosystems and complex interactions—new leaves provided new opportunities for herbivorous insects, which became food for carnivorous insects, and thus for mammals. The last common ancestor of placental mammals, in fact, had five functional copies of a gene that regulates the creation of enzymes that break down chitin, which is what insect exoskeletons are made of. Many Paleocene species have this inheritance, ready to make the most of the invertebrate buffet being presented to them. And many of those mammals are able to make quick work of the bugs thanks to a special form of shearing, crushing tooth.

Many of the surviving mammals, particularly the placental mammals, have what's called a tribosphenic molar. That's a mouthful of a term, but its expression is quite simple. As you might guess from the prefix, a tribosphenic molar is a cheek or chewing tooth that has three peaks. (If you want to get technical, these are called the paracone, protocone, and metacone.) The basic shape, which evolved in mammals during the Mesozoic, has a shearing part and a crushing part, which opens up a variety of different dining possibilities to mammals with this tooth shape.

That's different from other ancient mammals. The multituberculates like little *Mesodma* in her burrow, for example, had very complex molars. Her teeth were very good for crushing and grinding but not so good for cutting or shearing—something useful in chomping on leaves or muscle. Other ancient mammals had something called a symmetrodont molar, which is good for shearing but not so good for crushing. Now, other tooth types could have evolved from almost any of these, given the right evolutionary nudges. But much like why birds could not easily reclaim the throne of *T. rex* or crocs could not immediately start stalking the land again, having a single-use molar requires extra steps to take on other roles. Mammals with slicing molars might have enough variation that some are a little better at crushing food, and that trait could be selected for over time, but that's a long period shift compared with mammals that can both shear and crush as they please. The fact that many of the surviving mammals from the Mesozoic, now proliferating in the Paleocene, have tribosphenic molars means that there is automatically a greater range in their tooth shapes, behaviors, diets, and other food-related choices. Some mammals evolved to get better at shearing flesh. Others sliced leaves. Others crushed seeds and other hard foods. Some had a little of everything. Not to mention that certain tooth shapes relate to different ways of chewing. Mammals, unlike most non-avian dinosaurs, can chew their food and break it down to very small pieces before swallowing. This required changes to their tooth shape, their jaws, and their muscles, going back to the Late Cretaceous. In particular, some of these tri-cusped mammals can move their jaws in such a way as to grind—an essential ability for being able to pulverize plant foods, allowing herbivores and omnivores a more efficient way to get their nutrition, and therefore opening up some possibilities for bigger bodies. Even though the ancestors of Paleocene mammals survived by luck, the traits they carried with them were powerful and led to a spate of possibilities that would have been much more difficult, if not impossible, for dinosaurs to replicate. An ancient trait, one that had already been around for millions of years, gave Paleocene mammals the edge they needed. So equipped, mammals were poised to make the most of a riot of growth in forests unlike any the world had seen before.

Up until the evolution of angiosperms, the way plants reproduced largely rested on luck. Conifers would release pollen into the air, and if the winds were fortuitous, some of that pollen from a particular tree would meet the female cone of another tree, fertilize it, and develop seeds, and then those seeds would grow. The plants themselves were not so much active participants in the process but reliant on the luck of the draw. But not very long after their origin, angiosperms started to do something different. Angiosperms could effectively trick animals—particularly insects—into doing some of their dirty work for them. Perhaps that's not quite the right expression. The plants had no knowledge that they were doing this, no master plan to hijack the lives of social insects and fluttering moths. All the same, an evolutionary partnership between plants and insects began to develop. Flowering plants would do exactly that—flower—and insects would be drawn to those flowers, where both male and female gametes are held. Angiosperms that had a particular color or scent, purely by accident, might be more attractive. And so insects would visit those plants more, making it more likely that slightly more colorful, slightly more smelly plants would pass on their traits to the next generation—more so than the plainer ones. Entirely by accident, plants began to advertise. And this opened up plenty of new possibilities for insects: drawn to the colors and scents, insects would think that they were getting a nectar meal or some other resource, only to pick up pollen in the process and fertilize the next plant of the same species they stopped at.

Fruit has a role to play here, too. Fruit is another evolutionary happenstance that took off. Fruit was originally meant to protect seeds and provide them with just a little bit of energy in the form of sugars when they reached the ground. But insects, and eventually animals, began to take interest in these, too. Insects could use fruit to grow their young, just like any other part of a plant. Mammals and birds, on the other hand, ate palatable fruit, the acid-etched seeds deposited in some fertilizer after passing through. New ecological relationships began to evolve. In fact, some plants began to benefit from this sort of predation, so those with sweeter, less irritating fruits began to proliferate (although there were still plenty that kept

their toxic or unpleasant nature, no thanks to help from animal assistants).

Over time, there was so much more to eat. There were new
plants, bearing a variety of fruits never seen before. There were
new insect species that had their lives tied to those of plants,
nutritious resources in their own right. Diversity spawned
more diversity, each species providing a little push and pull on
the others in their habitat.

No species is an island. No species is a discrete and complete package by itself. A species is an expression of the interaction between organisms in its environment. Rather than being
boxed in by taxonomy, a species represents a point of interaction
between organisms. What would an *Eoconodon* be without the
bacteria in its gut or the parasites that live within its fur? What
would it be without the flesh it consumes from soft-shell turtles or the bitter berries it nevertheless can't stop munching or
the shade of the trees towering overhead and the bacteria and
insects that break down its feces? Any organism is a node that
is bound to and reliant on the others around it, whether there
is any direct interaction between them or not. The actions of
even one organism affect others, which affect others still, entire unseen webs of possibility that pulsate through each vital
moment—the thread of life itself.

For mammals to thrive, Paleocene forests had to shoot up.
There is more habitat in a forest than on an open, fern-covered
plain. Think of it this way: in a meadow dominated by ferns
and low-lying plants, there are only so many ways a mammal
can live. Mammals can burrow. They can hunt insects. They
can get big and graze on plants. But there is nothing to climb,

there are no fruits to eat, no trees to make nests in. An open area effectively creates habitat at three levels—underground, the surface of the ground, and, eventually, when bats arrive on the scene, the air. In a forest, however, there are many different levels. The mammals that can dwell in the canopy will not be the same as those that prefer the lower branches, which will be different from those on the ground, which will be different from those that burrow. Even on the same level—such as on the ground—there is a greater array of foods. There are the low-lying plants, fruit and seeds dropped from above, and, naturally, other mammals. These ecosystems become more complex, more intertwined, winding around each other. A forest opens up more opportunities to move in different ways—dig, leap, climb, run, glide—that more open habitats do not. This is the adaptive crucible that had to come first for mammals to undergo their second great evolutionary act, to stage a recovery and fundamentally change. And, strangely, more closed, forested places allow for a broader range of body sizes to evolve. The resources grow thicker here. Higher-energy resources—such as insects and fruits—are more common here than in open areas. That's why many open-habitat species are relatively small. Big species can keep them open, but these are usually grazers that can afford to eat mouthful after mouthful of low-quality, low-energy forage because of the way physiology changes with size. In terms of ways of eating, moving, and carving out a living, forests generate more diversity and disparity from within.

The shape of the forest guides the shape of mammalian evolution. Warm temperatures and frequent rainfall allow forests to grow throughout the year. That makes for multiple habitats, creating opportunities and interactions that would not otherwise be possible. More individual animals can live in a tropical forest than in the equivalent area of a grassland.

Such dramatic changes are not instantaneous. They take time to develop—millions of years, in fact. Specialization often comes from interaction and an ecological squeeze. Individual mammals are competing with other mammals for the same type of tasty plant. They can keep trying to compete and get more and more specialized—perhaps being better at sniffing out the food source, digesting it, breaking down its defenses—or they can shift to something else, another food source, or even be a generalist, taking a little of column A and column B. There is no right answer. There is only survival. All of this is to say that these early Paleocene mammals are not very specialized at any one thing yet. They are not like the dinosaurs in gaining extreme adaptations for self-defense, for crunching bone, for running fast, for climbing with agility. Those traits and adaptations can only come with time, from evolutionary unfolding that will open new possibilities and add new constraints. Someday there will be beasts that primarily, if not exclusively, eat fruits. There will be gliding and flying mammals capable of moving between patches of forest with ease. There will be granivores that make their living from gluts of seeds. There will be carnivores whose limbs are evolved to chase or catch, with teeth to puncture and crush with ease. But just not yet. At this time, all of these roles are still just distant possibilities, distant points of potential that depend on the individual moments of each life.

Somewhere in the forest beyond, a bird lets out a call that echoes like a sharp report through the forest. It is not one of the nocturnal birds that forage by darkness. It's the opening of the dawn chorus, a cacophony of sound that rouses *Purgatorius* on all but the most exhausted of mornings. She nestles closer against her little ones as the sky begins to brighten, the orange morning sun—bright as the tasty yolk of an egg—begins to peek above the distant horizon. For a moment, little more than a passing second, the low angle is just right to send a beam of sunlight through the dense forest to her tree. The orange light washes over *Purgatorius* and her offspring as it passes. She blinks at the light and watches the shape of the shadows made sharp by the morning light. It's a morning much like the one that came the day before, and the one that would come after, the same and still shining on all the subtle differences that had passed in the previous day. Some were born. Some had perished. Some were injured. Some grew. Life, resilient and strange life, continued, the beginning of another day in the new Age of Mammals.

Conclusion

66.043 Million Years After Impact

"Hey, do you want to come with me to look at some rocks?"

I'll admit, it wasn't my best pitch. I may have even under-
cut the significance of what I was proposing to Splash, my
equally fossil-focused girlfriend, just because I know I can always make her laugh through understatement. Of course she'd
want to come. I was going to the line in the stone where the
Age of the Dinosaurs ended, ten hours to the southeast by car.

"We can leave on Friday night, get to Grand Junction, then

drive six hours the next day to the spot and back," I said, feeling a little less sure about asking my partner to endure hours of sitting still while zipping along the winding canyon roads of Colorado. But I needn't have worried. A few days later, with Jet, my German shepherd, curled up in the backseat, we three Cenozoic mammals struck out for the K-Pg boundary.

The spot isn't difficult to find. Even though there were technically closer sections I could have visited, such as the Late Cretaceous and Early Paleocene strata in the North Horn mountains of central Utah, the season was getting late and I didn't want to lead a wild-goose chase onto mucky mountain back roads where traffic jams of free-range cattle are common. A long swath of the K-Pg boundary is exposed in Long's Canyon Trail, near Trinidad, Colorado. It even pops up in Google Maps—type the outdated title "K-T boundary" into the app and it shows up immediately.

I had to go. I needed to touch that stone. I didn't feel right, spending so much time imagining the world of the Cretaceous and the days after without putting my finger directly on the point where one world of possibilities was transformed into another, a date that I owed my very existence to. Seeing sections of the K-Pg boundary in museums—cleaned and safe behind exhibit glass—just didn't feel like enough. I didn't want this dramatic event to live only as an idea in my head or a body of technical papers sitting on my hard drive. Science is a way of knowing, but hardly the only way. I wanted to know what I might learn from reaching out to touch the wreck of the Mesozoic world.

I sometimes say that I miss dinosaurs. A close friend of mine often calls me out on this. How can I miss creatures that I never got to see alive? How can I miss a time that I never got to experience other than as an amalgamation of museum displays, paleoart, research papers, and fill-in-the-gap guesswork that lives in my head? There's actually a term for this—*anemoia*—but it's most often used for people who are infatuated with their idea of the 1960s or who desperately wish they could have seen Queen with Freddie Mercury as front man. The term isn't usually used to refer to someone like me, prone to daydreaming about forty-foot monsters clad in pebbly scales with their fuzz softly wafting in the Late Cretaceous breeze. Still, from the time I was very small, I connected with the terrible lizards. Their skeletons, often cobbled together from multiple individuals and framed with anatomical inaccuracies—dragging tails, the wrong skull—were my friends. My love was unconditional. You can't be hurt by a friend who's extinct. I wanted to know them, to see them in their prehistoric heyday. I still do. Even to watch a fuzzy, beaked *Thescelosaurus* dozing in the shade of a dawn redwood or an *Edmontosaurus* chewing through a Cretaceous meadow would be an incomparable joy. Every organism is an absolute marvel of evolution worthy of our awe, down to the simplest microorganism, but there are those species we connect to more strongly than others. And for me, that pull is incredibly strong to the world of 235 to 66 million years ago.

I'm sad that the non-avian dinosaurs are gone. Sometimes I can't quite believe it. The phrase "when dinosaurs ruled the Earth" is cliché, easy to take for granted, but there's at least a speck of truth there. For tens upon tens of millions of years, dinosaurs thrived in such unabashed numbers that we still haven't found them all yet. Even at the present rate of one new dinosaur species named every two weeks, we are truly only just beginning to find them, much less understand anything about them. The relatives of the great *Torosaurus* and *Triceratops* are wonderful examples. Even though these dinosaurs—called ceratopsids—shared the same basic body plan, their ornamentation was so wildly different that they're easy to tell apart at a glance. Brow horns, cheek horns, nose horns, rough bosses of bone tissue, spikes, hooks, and splashes of little horns around the frill—each a dinosaurian fashion statement represented by dozens of species, with new variations found and described each year. These animals were not oddities or rare beings. The world was theirs much more than it's ours. But they're not here. Every last lineage, save for the birds, perished. For all that dinosaurs survived, adapted to, and endured, there are simply some forces of nature that can't be argued with, avoided, or escaped. Dinosaurs were lucky, but good fortune is so often balanced by bad.

Had the non-avian dinosaurs survived, however, I wouldn't be here to tell you these stories. If there is any clear message in the fossil record, any distilled takeaway that the entire history of life on Earth speaks to with a full and clear voice, it's that no species is inevitable. Every organism lives where accident and circumstance of life meet each other, variations providing the raw material for natural selection and other evolutionary forces to shunt down different pathways. Not that there is any intent to this. It's a passive state, a constantly running routine that is merely part of existence itself. It's a difficult fact to grasp. Our existence, our origin, relies on an unbroken string of happenstances and accidents, each just one possibility among many. This is another expression of the famous butterfly effect, an idea first explored by the science fiction impresario Ray Bradbury in *A Sound of Thunder,* a story about people who go back to the dinosaurian prime only to unravel the future because one person accidentally steps on a particular butterfly.

But we're not talking about a lone butterfly. We're talking about an unprecedented impact unlike any that came before and unlike any that has occurred since. Had the asteroid struck a different part of the planet, or missed entirely, the devastation of 66 million years ago would have been altered, at minimum, if not avoided altogether. Even in a lessened mass extinction, perhaps killing half of all non-avian dinosaur species, there's little reason to think that the saurians wouldn't have bounced right back, spinning off new forms that would be strutting around this planet this very second in an alternate history where the Age of Dinosaurs never came to a close. Mammals would likely still be around, too—just as they had been throughout the dinosaurian heyday—but their evolutionary options would still be constrained by sharing the world with dinosaurs. Perhaps there would still be primate descendants of little *Purgatorius* running around forest canopies and grasslands, but the sequence of unexpected events that led to our evolution would have been unintentionally ushered down a different pathway of possibilities. There wouldn't have been a way to get to here from there, and there likely wouldn't be a primate self-reflective enough to even notice that such is the nature of time itself. We're riding along time's arrow and beholden to its restrictions just as the dinosaurs were.

Had the non-avian dinosaurs survived, I thought to myself as I drove off the pavement and onto dusty dirt roads leading to the trailhead, then none of us would be here to miss them. And if the life I knew hadn't collapsed two years prior,

I likely wouldn't exist, either. I had only been able to come out as transgender in the wake of my divorce—years of my life turned to ash as something long dormant started to take root. Painful, dangerous, and scarring as it was, I needed the life I knew to end. I could miss parts of it, but who I really am and who I tried to be could not exist in the same reality. One had to cede to the other. Windows down, warm October air flowing through the car, the slap bass of the Red Hot Chili Peppers' "Give It Away Now" thumping through the car speakers, Splash, Jet, and I rolled up to the sign marking the hike. I had to laugh at myself. The posting read

LONGS CANYON
SELF-GUIDED TRAIL

. . .

BOUNDARY
.25 miles

I had driven a total of ten hours on an overnight trip to walk a quarter of a mile. But this is what I had been waiting for, what I had been fixated on as we rolled through all those miles of asphalt. I let Jet off his leash and our little mammalian family wandered up to the line.

Even if you're not a geologist, the strike through the Cretaceous stone is impossible to miss. Above, the stone is a buffed and weathered tan, sagebrush and other shrubs anchored into it. Below, the slope is powder gray, the pulverized remainders of the soot-rich layers of rock. And right in the middle, there lies an unmistakable dividing line. Thick, flaky blocks of clay-rich rock are held close by darker bands just above and below. If I were able to take samples of each part through the sequence, that clay-filled middle part would show a spike in iridium, the mashed-up and microscopic bits of the deadly asteroid that rained down through the atmosphere. I don't know if I've ever seen a more beautiful section of strata in all my time wandering through the western deserts.

I walked up to the rocky overhang, already well shaded so late in the day, as Jet wagged, bounded, and ran up to my sitting spot while Splash got happily lost in her own contemplations. I reached out and pressed my palm to the boundary layer. I didn't expect to have an epiphany. I didn't hear any voice in my head except my own. I stumbled as I tried to find the right words. What should I say, now that I'm here? I owed my existence to this disaster, but it's one that I might have elected to cancel if I had the ability to. I tried to clear my head as I shifted in my seat of ancient, sooty stone. This place has no equal. We haven't invented a category for what such a monument represents. It's both death and birth, the end and the beginning, an inflection point that carries continuity through the middle. I'd come not only to visit a gravestone but also to see the place that allowed me—that allowed all of us—to exist.

After a time, Splash, Jet, and I wandered off. We'd driven so far, and we'd have to go six hours back in just a little while. Might as well stretch our legs while we could. We all walked along the trail a little farther, stopping at a wildlife blind to watch a gaggle of waterfowl in a pond below. Jet jumped up to peer through the viewing ports, too, perhaps confused as to why we were watching the birds rather than chasing them. But we'd leave the dinosaurs be. They are the recipients of bad fortune and good just like we are. Birds evolved during the Jurassic, and thrived during the Cretaceous, but their history would have been altered by a canceled extinction, too. Every bird—from penguins clutching Antarctic ice between their toes to the ducks paddling around that Colorado sanctuary—is a dinosaur that evolved in a post-impact world, their history molded by the event just as ours was.

Perhaps the strangest aspect of this entire tale, the collapse of the Mesozoic, is that it didn't happen sooner. In endeavoring to understand the how and why of the end-Cretaceous extinction, paleontologists have discovered that life in the terrestrial realm has always been relatively unstable. Experts count this as species turnover—how long a particular species is around and whether it is replaced by a similar species carrying out a comparable ecological role. In the case of the species that lived during the last 10 million years of the Cretaceous and the first 10 million years of the Paleogene, creatures that lived on land don't stick around very long. Terrestrial species tend to arise and disappear every 1 to 2 million years. But that's not true for the semi-aquatic species. Animals like the soft-shell turtles and crocodiles stuck around for several million years, perhaps 10 million or more. You could go to a swamp from the time of *Tyrannosaurus* and see more or less the same species you would if you hopped back another 10 million years into the past.

No one knows why. Freshwater habitats seem to offer a kind of buffer; or they are habitats where evolutionary turn-over happens more slowly. The creatures that dwell in ponds, streams, lakes, and swamps are not as drastically affected by changes to climate, even though many of them are cold-blooded. There's some insulating, stabilizing factor that hasn't been uncovered. But what is clear is that creatures on land have a much more tenuous existence. They go extinct at a much more rapid rate, either disappearing entirely or ceding ground to their descendent species. Consider this. The crocodile *Brachychampsa* is found in 80-million-year-old rocks of the American West, all the way through the end of the Cretaceous 66 million years ago. Now think of *Tyrannosaurus*, a dinosaur that lived between 68 million and 66 million years ago. If each *Tyrannosaurus* ancestor along the way was around for that long, that's seven different species that came and went as *Brachychampsa* watched from the weed-choked shallows.

None of this means that the non-avian dinosaurs, or any other Cretaceous creatures, were just waiting for extinction. Nor that they somehow deserved extinction or were becoming outmoded. The upshot is that terrestrial ecosystems in the Late Cretaceous were very sensitive to perturbations. The constantly shifting ground—in climate, in how species interacted on the landscape, in vegetation—drove a faster rate of evolutionary change. Add a seven-mile-wide asteroid to the mix and you get a rapid shake-up that most organisms cannot cope with. The creatures in the water or those that can burrow get a reprieve. Those on land with more specific requirements, however, did not stand a chance—they might not have even if the effects of the impact had been somewhat gentler. The impact and its effect underscores the fragile construction of terrestrial ecosystems. It's not just a once-in-an-epoch event;

rather, it emphasizes how complex ecosystems can be rapidly torn to shreds because the instability is part of what generates so much evolutionary beauty.

But understanding all these things will not bring our beloved dinosaurs back. We can make sense of the catastrophe, and even replay some of its moments, but the world as it was is gone. All we can do is dig a little more, wonder a little more, and stretch our imaginations to restore a world we'll never be able to see. Genes don't have the longevity to ever construct a *Jurassic Park,* and claims that something like *Acheroraptor* could—or should—be engineered by modifying a chicken is just fantasy. If you want to see or touch a non-avian dinosaur, you have to go where their bones are kept.

In such moments, when I can find a quiet place in a museum hall, often near the end of the day when footsore visitors are headed toward the gift shop, I wonder why I ache for these creatures the way I do. Why any of us do. The K-Pg disaster was the fifth mass extinction on the planet. We don't shed bitter tears for the trilobites that disappeared at the end of the Permian or the jawless fish that were almost entirely eliminated in the Devonian. I can't think of an answer beyond love. Even if we're not certain why—is it their impressive size, their bizarre ornaments, their supposedly rapacious habits?—we hold our dinosaurs dear. Often, the creatures are our first encounters with big ideas that frame life on this planet. This is a *Triceratops.* This majestic herbivore lived a long, long time ago. What happened to them? They went extinct. An asteroid struck the planet, and then there weren't any more *Triceratops.*

Dinosaurs continually spark our curiosity. I truly believe this is why we venerate them in museums and love to watch them tear across movie screens. We don't have any appropriate metric for such creatures in our modern world, especially not during our own megafaunal lull when there are hardly any giants outside the whales of the sea. To look at a dinosaur is to wonder what that animal was like in life, what that world was like. The world was so full of them, and for so very long, that we don't wish to accept that such creatures could so rapidly disappear, that life could take such a quick and cruel turn. Perhaps this is a little bit of projection onto the past. If dinosaurs could rule the world for so very long, more than twenty-eight times longer than humans in any form have existed and more than twice as long as primates, then perhaps they are the ultimate memento mori. We gaze at the grinning skull of a *Tyrannosaurus,* light glinting off those permineralized teeth, and we see what may become of us simply by chance—if not by our own hand.

Imagining a dinosaur is an act of raising the dead. We do it in film, in art, in museums, and in books like this one. It's an ability that's unique to us in the present world, an understanding not just of the past but how the past flows into the present. And it offers us a challenge. Dinosaurs live again where our imagination touches bone, the consequences of impact creating a great, constantly unfolding puzzle in which the discovery of every new fossil feels like a victory. Against the odds, this creature was fossilized. And against the odds, we found it. The immense distance of time between us and the last day of the Cretaceous has created a lost world to explore, places where even familiar ground can hold mysteries. Any geologic map of fossil-bearing formations may as well say "Here Be Dragons" across the bottom.

Perhaps it's that conflict that keeps me going back. The Age of Dinosaurs is gone forever, but we have enough pieces to touch it by proxy. That time, those creatures, can never be restored, and yet I wish to know them. I'm curious. I think many people are. And curiosity doesn't like to sit still. I can logically understand why the Cretaceous came to a close, and how those events are essential for my own existence, but logic rarely wins out against emotion, and healing from such losses is always incredibly difficult.

I don't know whether the K-Pg disaster, and the sixth extinction crisis we're hurtling toward, holds any lessons for our modern world. During the grip of the cold war, discoveries of just how severe the impact winter was underscored just how awful nuclear winter could be if the United States and the Soviet Union tried mutually assured destruction. The world had been through such a disaster once. Best not to repeat it. And knowing what is possible—what has already happened once— NASA monitors asteroids to get a bead on whether any might strike the planet. That's a wise decision, certainly, especially given how we continue to spread out over the Earth. Even a localized impact could be a terrible disaster. But it would feel trite to use an entirely accidental and unavoidable event as a lesson, as if it were a mistake or something under our control.

There may not be anything that transpired between 66 million and 65 million years ago that can make life better today or that can make us wiser. Much as many of us might have wished to be one when we were kids or when we're playing video games, we are not dinosaurs. The threats we face are different, and often self-inflicted. We have known the reality of human-caused climate change for decades and have done shockingly little to avoid its consequences. But perhaps this is all a symptom of looking in the wrong place, or hoping for a distilled nugget of advice that gleams as beautifully as polished amber. In all the time I've spent pondering this event, of trying to think my way through the tangle of an ecosystem I'll never visit, of poring over strata sections behind Plexiglas and getting my boots right on this terrible, wonderful, transcendent line in the stone, what has struck me is life's persistent, vibrant, and unabashed resilience.

From the time life first originated on our planet over 3.6 billion years ago, it has never been extinguished. Think about that for a moment. Think through all those eons. The changing climates, from hothouse to snowball and back again. Continents swirled and bumped and ground into each other. The great die-offs from too much oxygen, too little oxygen, volcanoes billowing out unimaginable quantities of gas and ash, seas spilling over continents and then drying up, forests growing and dying according to ecological cycles that take millennia, meteorite and asteroid strikes, mountains rising only to be ground down and pushed up anew, oceans replacing floodplains replacing deserts replacing oceans, on and on, every day, for *billions* of years. And still life endures.

Life can be gravely injured. Individually, or even as a species, lives end. Life can be constrained, cut back, forced into uncomfortable shapes, and stressed so gravely that the trauma is visible in fossil beds as surely as the tissues of a living animal. But life is still here. Vibrant, resilient, overwhelming life. Life is not static. It is responsive. Have you ever walked down the sidewalk and seen where a tree has grown into the criss-cross pattern of a chain-link fence, or pushed up the concrete sidewalk slab with its roots? That is what life does, day and night, as long as it exists.

I have no doubt that the years I've spent as a transgender woman have affected my perspective. How could they not? There is a natural reality that we all live in, but how I perceive it is altered as much by my identity as by the fact that there are some colors of the spectrum I cannot see, some sounds I cannot hear, some sensations I cannot feel. The whole of our story is not the whole of life's story. Nevertheless, when I think about these sweeping changes, it's difficult for me to focus on the losses. My personal perturbations and crises have taught me to celebrate the growth.

I wrote this story as my personal topography, figuratively and literally, veered off the course I expected. Parts of my life, my body, have risen and fallen, altered shape, and changed function, in a way that makes me think of the waltz of continents and the shape of the deserts where ancient worlds still lay, as yet unknown to us. The history, and the record of that history, show a responsiveness, an endless interplay between luck and biological foundations for change. I lived my life one way for what seems like eons, my personal Mesozoic era. In the aftermath of personal tragedy, when the constant burn left me feeling that everything I knew had been destroyed, small reminders—little things that had always been there— suddenly became more important, the small, seeking shoots rising up through the wreck, thirsty for each new day. My life wouldn't be the same, couldn't be the same anymore, but all those small, precious things—in my heart, waiting in my body—would finally have room to grow. That freedom could not have existed without the searing pain and the end of an era, just as we could not have come to be in a world held in the claws of the terrible lizards. Beginnings need endings, a lesson that we can either hold carefully or that we can deny until it finds us.

Just consider what life is. Biologists of all stripes have been pondering, debating, and arguing over this point for as long as there have been biologists. The search is often for a common denominator that is unassailable and airtight, that can be applied as a litmus test to divide the living from the nonliving. Perhaps we might even be rewarded for distilling such a pure truth about nature with a further secret, that we will gain something powerful as a result. But I've never been much for reductionism. I am not so much interested in what life *is* as what it does. Life endures. Life changes. Life is a beautiful and horrible thread made of uncountable organisms, a relationship that can do virtually anything except remain the same.

Perhaps that's oversentimental for a science book. This is not a tome of Cretaceous poetry. But it's my small, grasping, perhaps failing attempt to pay tribute to something just beyond the reach of tangibility, something that is there and yet so deep that I don't know if we will ever see the bottom of it. It's on my mind when I watch a quail nervously bobble down the sidewalk, when I see the peach tree in my yard start to put out new fruit for the season, when I stop to watch a jumping spider do its jerky little dance on the rough surface of a telephone pole. All of these things, wild or domestic, familiar or strange, are connected to me, not just now but through history. They all have a fossil record. They all leave some mark on the world around them. They all came from something, new and amazing expressions of a world that was founded the moment that asteroid touched down on our planet's surface and altered the possibilities for life on Earth. If you're having trouble seeing it, consider it this way. We are now 66 million years removed from the K-Pg disaster. That is a very long time. But 66 million years before the K-Pg impact itself, flowering plants were new, tyrannosaurs were tiny, many birds still had teeth, and the Age of Dinosaurs was in the middle of its seemingly endless summer. We, today, still live in a world where we can start at almost any point and see how the world's fifth mass extinction influenced and altered that point. It's not just the great scar of the Chicxulub crater that remains. It's the splash and paddle of the duck-billed platypus, the flowering of a magnolia tree, the bee getting pollen-drunk in the garden, the can of beans in the supermarket, the robin scrounging for worms in the grass, our forward-facing eyes and our grasping hands—the uncounted, stunning facets of life that owe something to unforeseen disaster that met us by happenstance from across the cold expanse of space.

I would love to button up this thought with one of my favorite film quotes, delivered perfectly in every dinosaur fan's number one favorite—"Life, uh, finds a way." But I can't do that, and not just because it feels a touch too trite. Life has no plan or direction. Life is not seeking a way and burrowing into those cracks and crevices with any thought of tomorrow, or even the next moment. But I think, pretentious as he could be, the fictional Dr. Malcolm was sidling up to a slightly different formulation, one that includes constraint and possibility, growth and death. And it is simply this. It's the lesson that those million years between the Age of Dinosaurs and Age of Mammals teaches us.

If there is a way, life will find it.

Appendix

I wish I could say that I spent days, weeks, and months in the Cretaceous and Paleocene, carefully observing and taking notes on the behavior and habits of life in that precious one-million-year window when life came back. Sadly—and I say this knowing that such an adventure would likely lead to my demise, by the K-Pg extinction fallout if not at the claws of the hungrier dinosaurs—I did not actually venture back more than 65 million years to dig up the components of this book. Time travel exists only in the sense that we're moving through time, passengers on time's arrow, and so I'll never get to see *Tyrannosaurus, Eoconodon,* or any of my other favorites in life. Instead, what you're holding in your hands has been stitched together from various scientific sources covering everything from the effects of wildfire on a forest to the pebbly texture of *Triceratops* skin.

Naturally, I felt that dipping in and out of the K-Pg narrative I set up might prove jarring. It'd be like watching *Jurassic Park* but with a mandatory paleontological commentary, pointing out each fact and shred of speculation. I love this approach when touring with friends through museum halls, chattering away about how we know what we know, but it doesn't fit so comfortably when I'm asking you to place your mind-set back tens of millions of years in the past. "*Tyrannosaurus* scans the clearing, but she has nothing to fear. She has not seen another *rex* in days, and these hulking dinosaurs maintain large territories—a fact attested to by hypothetical tyrannosaur counts in Marshall et al. 2021 and Paul Colinvaux's classic ecological study *Why Big Fierce Animals Are Rare*." See? So instead of citing you fossiliferous chapter and verse for each point, I've created a chapter-by-chapter appendix to explain what we know, what's hypothetical, and how much speculation I used to smooth over the gaps where research has not yet given us the answers we desperately wish to know about long-extinct creatures.

On that note, I expect some readers might be a little surprised, even confused, why I've selected the impact of a huge asteroid as the primary trigger responsible for the K-Pg extinction. Paleontologists have argued about this point, perhaps most famously in 2010. In that year a virtual battalion of geologists published a paper in *Science* affirming that the world's fifth mass extinction was unquestionably caused by the bolide impact. Even though experts had proposed scads of ideas for the extinction and the more specific demise of the non-avian dinosaurs during the previous century, the authors brushed all those ideas aside—from ravenous caterpillars to the burbling lava output of the Deccan Traps—to confirm the impact as the prime reason the Cretaceous ended with a bang. But scientists do not move in lockstep, as if each new paper represents an unquestionable gem plucked from nature and forever ensconced as Truth. Other experts disagreed with the assessment in *Science*. A group of vertebrate paleontologists wrote in to say that the extinction was caused by multiple factors, from volcanic activity and sea level changes to the impact itself. There was no single smoking gun, in other words, but a combination of changes and disasters that became something greater. Still others ran to their word processors to say that the impact's effect was negligible and, in fact, the eruptions of the Deccan Traps were the main extinction triggers. And there was a third response, claiming that the original paper overlooked a vast field of data and that, instead, a combination of impacts, volcanic activity, and climate change played out over a very long time to winnow away life's diversity.

But science does not operate by vote, and the sheer number of yea or nay attestations doesn't verify or negate a hypothesis. In the case of a dramatic, disastrous event that we missed by 66 million years, we only have the remaining evidence and the ever-changing theoretical frameworks we use to understand the critical moment in life's history. In the years since the exchange played out in *Science* in 2010, researchers have continued to question, study, and refine. In the process, the effects of the asteroid impact have proved to be worse than previously thought—while some of the other suspects have either been exonerated or had their roles recast. Research into the massive eruptions of the Deccan Traps, for example, has revealed that the outpourings of lava and greenhouse gases had little negative effect on the world's biota. In fact, the Deccan Traps may have helped spare some of the organisms that were hanging on during the earliest days of the Paleocene—the greenhouse gases spewed into the air by the volcanoes keeping the atmosphere warmer than it otherwise would have been, skirting along the edge of a potentially deadlier deep freeze. Likewise, while sea levels were dropping during the end of the Cretaceous and the world was becoming cooler, dinosaurs and other forms of Mesozoic life had survived these changes before. Even if these shifts caused the extinction of some species or the extirpation of a population from an area, they were insufficient to cause a global, near-instantaneous extinction that can be seen in the rock record. Perhaps the fact that the previous four mass extinctions and lesser extinction crises happened over hundreds of thousands of years led paleontologists to expect that the same would be true of the fifth time around, but that doesn't seem to be the case. The K-Pg mass extinction is entirely unprecedented in Earth's history, and that's precisely why I became fascinated with the catastrophe. This was not a grinding, painful change caused by belching volcanoes or acidified seas. Death came suddenly and took its toll quickly, creating conditions that life never evolved to cope with. Despite all this, life on Earth persists.

I decided not to hedge on the cause of the K-Pg mass extinction. I did not set out to tell this story in a way that would please everyone or only communicate the science as written. That would please no one. I feel confident that the impact of Chicxulub 66 million years ago sparked a global crisis unlike any seen before on this planet; the combination of incredible post-impact heat and an extended impact winter explains the pattern of extinction from the last day of the Cretaceous to a million years into the Paleocene. In some cases, perhaps, there were species that went extinct during this same period because of more mundane causes—like bivalves that lived in the remnants of the Western Interior Seaway, only to have their habitat dry up as sea levels changed. I feel that's entirely plausible, even likely, but such small-scale events are not what we're talking about when we focus on the K-Pg extinction. Discussions of this event have always focused on the big picture—on what happened to end the Age of Reptiles and so many other forms of life that had come to thrive during the Triassic, Jurassic, and Cretaceous. Ultimately, extinction is the fate of all species, certainly, but when so many players exit the stage at once, it's a sign that there is something greater going on than another expected scene change.

Naturally, the loss of the non-avian dinosaurs, the pterosaurs, the ammonites, and so many other forms of life really forms only the background of this story. The story I've endeavored to express is not so much a tale of deeply felt loss as it is of renewal and recovery. Even though plenty of ink has been spilled about the K-Pg mass extinction—what caused it, who perished, and how scientists fought tooth and claw about it— little has been said about life's recovery during the Paleocene. That's a shame. The rise of the mammals during the Cenozoic was not inevitable by any means. After all, avian dinosaurs survived. Some got very big and fed on smaller mammals. Members of the crocodile family also passed through the event and some evolved into menacing terrestrial hunters. There could very well have been a second Age of Reptiles, but why that didn't happen—why mammals became such an ecological centerpiece—is a story that requires us to sift through the aftermath. Had history played out just a little differently, even when the worst of the extinction was over, it's entirely possible that we never would have evolved and mammals would have remained relatively small, snuffling creatures that adapted to cope with a world held in a sharp-taloned reptilian grip. There is far more to this epic than the suffering of the terrible lizards, and I endeavored as much as possible to highlight the meek organisms that held on and eventually thrived as the dinosaurs fell.

What follows is a chapter-by-chapter breakdown of our players and what we know about them. And, really, if I'm going to give myself all these creatures to play with, I'm going to have my speculative fun, too. It's part of a long tradition in paleontology, and, truthfully, the combination of fossilized fact and informed fiction is present in every museum exhibit or piece of art you've ever seen about the Age of Dinosaurs. I was struck by this while indulging in a little dinosaur time in Denver, Colorado, where an exhibit highlighting Sue the *T. rex* was in town. When my girlfriend and I walked through the timed-ticket entry doors and into the dark of the exhibit, we were immediately met by the gaze of a life-sized Sue proudly carrying a small *Edmontosaurus* in their mouth. The tyrant was an impressive sight, seeming incredibly bulky yet regal with shiny plates of keratin covering the ornaments of their skull. Sue did not have any fuzz on their frame, much to my disappointment, but I wouldn't hold it against the artists behind the sculpture—the facsimile felt like it had weight from another time, perhaps the closest I would ever be to seeing a tyrannosaur in the flesh. But, I thought, I had felt that way before when I visited the exhibit of *T. rex: The Ultimate Predator* in New York the year before. The museum's version of *T. rex* was bright orange, with a mane of bristly protofeathers, and the skin of the arms was so thin as to make the appendages look very reedy. This *T. rex* had its jaws agape, drool cascading between banana-sized teeth, even though—from just the right angle—the whole thing had the charming appearance of a *Muppet Show* monster. The inspiration was the same— the bones of an animal that we've known better than most dinosaurs for more than a century now. Yet two different institutions, working with two different teams of artists, came up with visions of the same animal that looked like they could very well be different species. Each group of researchers made their own decisions and artistic choices, leading to strikingly different views of the most famous dinosaur.

To say that we must stick strictly to the science doesn't make a lick of sense. Science is a process, not an answer book. Fact and theory are intertwined, and if we were to express only the unvarnished data we wouldn't be left with much to say at all. An illustration from a very old paleo book, *The World Before the Deluge,* helps make the point. When the book was published, in 1863, paleontologists were still aflutter about a particular reptile from the Jurassic limestone of Germany. A fossil feather had been discovered in 1860, followed by a partial skeleton framed by delicate feather impressions in 1861. This was *Archaeopteryx,* the urvogel that acted as an evolutionary connecting point between reptiles and birds. There was just one problem: that first exquisite skeletal specimen had no head. Did *Archaeopteryx* have a beak? Teeth? Both? Neither? There was no way to answer the question, and so *The World Before the Deluge* portrayed *Archaeopteryx* flying high above Jurassic conifers totally headless.

In one sense, a headless *Archaeopteryx* could be called accurate. No head had been found, so any illustration of the animal's skull could be easily disproven. In fact, the skull of *Archaeopteryx* had already been found—it was just mistaken

for the head of a fish. But we also know that *Archaeopteryx* was
a vertebrate and, more than that, somewhere between reptile
and bird. Illustrating a headless *Archaeopteryx* could be defended from a scientific standpoint, but it simultaneously runs
counter to what we know about animals and their anatomy. In
my view, at least, it would have been better to take a gamble
on the Jurassic bird's hypothetical appearance than to draw a
bunch of disembodied bones and feathers in the air as if we
were awaiting a perfectly preserved *Archaeopteryx* study skin
to appear in the lithographic limestone quarries of Bavaria.
And, considering the time over 80 million years after that of
Archaeopteryx, I elected to fill in my expectations of the Cretaceous and Paleocene worlds on the basis of what we know and
what we expect. Some things I've written may very well prove
to be wrong. Actually, I'm quite certain that some will be, and
I'll be glad for that. To make a guess and find out you're in
error is an opportunity to learn. Not to mention that any view
of prehistoric life is a product of its time. This is how I see
and understand the end of the dinosaur era from my current
standpoint, still in the early years of the twenty-first century.
If we're to be honest with ourselves and recognize that prehistoric life requires imagination to fully envision, then we have
to accept that our imagination has its own influences and limits. If someone tried to write this same book ten, fifty, a hundred years earlier, the story and the creatures within would be
fundamentally different. Why should I expect to be free from
the same constraints?

When I was an elementary school student, I would often
be reminded by my math teachers to show my work. Getting
the right answer didn't matter if I didn't understand the pro-

cess. That's what I intend to do here—to show my work and offer a peek behind the paleofantasy into what we know and what we have yet to understand. What follows is the raw material that inspired me to think about the effects of acid rain on fossils-to-be, on how turtles could stay underwater through the worst of the heat, and what mammals of the early Paleocene forest might get up to. Even though we don't have that wished-for time machine, at least we have scraps of time to feed our churning imaginations.

THE DAY BEFORE IMPACT

In all of prehistory, there is no animal that commands our attention quite like *Tyrannosaurus rex*, the king of the tyrant lizards. Since the time this dinosaur was officially named in 1905, the enormous carnivore has stood as the ultimate dinosaur. Naturally, I couldn't very well write a dinosaur book without giving the Cretaceous celebrity some time in the limelight. The inspiration for our opening scene, however, didn't come directly from a paper or field site, but from an experience I had in Yellowstone National Park. During a visit to the park in 2017, I noticed a crowd of people gathering along a turnout in the park's Hayden Valley. A bison had died near the road, seemingly of natural causes. Such a bounty was sure to attract carnivores, and, sure enough, there were golden eagles, ravens, and other birds perching on the old bull's body. The trouble was that none of the birds had the ability to get through the bison hide. They had to wait for another carnivore to open the carcass for them. Lucky for them, a grizzly bear slunk out of the tree line and headed straight for the dead bison just as the valley slipped into darkness. The next day I watched as a grizzly—presumably the same one—and a gray wolf took turns gorging themselves with bison meat, the ravens and raptors hopping around to nab what tidbits they could. I wondered what it would be like if such a scene played out in the Hell Creek ecosystem of 66 million years ago.

To set the scene, I drew from two lines of evidence regarding the poor deceased *Triceratops*. The first involves the social lives of these animals. In the strata of Hell Creek and equivalent formations, *Triceratops* represents a paradox. Paleontologists have collected dozens of *Triceratops* specimens through the years—especially skulls—and a survey of one portion of the Hell Creek Formation found that *Triceratops* was the most abundant dinosaur on the landscape. If you were to wander around Hell Creek at the end of the Cretaceous, you'd be most likely to see the three-horned herbivore. Yet most *Triceratops* finds are isolated. There are no great boneyards of *Triceratops* containing dozens, even hundreds, of individuals as there are in the older rocks of 10 million years earlier. On top of that, one of the only *Triceratops* bone beds known contains young, subadult animals—a pattern seen among other non-avian dinosaurs. This might indicate that parent *Triceratops* did not care for their offspring for very long—apologies to fans of *The Land Before Time*—and so youngsters grouped together in order to better spot potential threats on the landscape and play the numbers game of predation, with life in a social group making it less likely that any particular individual would be singled out to be munched by a tyrannosaur. Perhaps adults formed herds as well, but then again *Triceratops* was large enough and well defended enough that they might not have needed to worry as much about being ambushed by a tyrannosaur, and the old bull in the story is a tribute to this idea.

How the bull perished also draws from recent research. Paleontologists have known for decades that dinosaurs got bone cancer. Most cases identified so far seem to have been benign or like they did not severely affect the health of the animal. But in 2020 researchers reported the first known case of a malignant tumor from the limb bone of a horned dinosaur that lived about 10 million years before the time of *Triceratops*. The bone of this horned dinosaur, known as *Centrosaurus*, came from a vast bone bed created when a coastal storm flooded a river, decimating a huge herd of these animals and scattering their bones across a broad floodplain. Given that bone cancers are part and parcel with being a vertebrate, part of a more ancient inheritance of bone itself and all the ways it can be injured, it's not a far stretch to expect that *Triceratops* might have had to deal with the same ailment as its older relative.

After death, the body of the *Triceratops* would have been difficult to break into for most small carnivores. A well-known specimen of *Triceratops*—albeit one that has not yet been formally described despite being on display for years—indicates that this dinosaur was largely covered in large, quarter-sized scales. Paleontologists have been finding scaly skin impressions of similar dinosaurs for well over a century, and despite the fact that *Triceratops* may have had some form of bristles or protofeathers based upon such ornamentation found on related dinosaurs, most of the dinosaur's body would have been covered in a thick protective hide. That would have presented a challenge to whatever scavengers happened upon the carcass—

the beaked pterosaurs and birds of Hell Creek did not have the serrated teeth or the jaw strength to break through the skin and probably would have picked at the more accessible soft parts, as scavenging birds do today.

That *Tyrannosaurus* would have been capable of rending open an adult *Triceratops* is attested to by the fossil record. While computer simulations have indicated that *Tyrannosaurus* certainly had enough jaw power to shatter bone, the verification comes from *Triceratops* fossils scored by tooth marks that can be attributed only to *T. rex*. A particular set of *Triceratops* hips—known to experts as MOR 799—had deep punctures that could only have been created by the bite of a large tyrannosaur. Several *Triceratops* skulls have even been found with a particular pattern of bite marks, indicating how *Tyrannosaurus* bit just here, snipped just there, to cut the heavy neck musculature of *Triceratops* and wrench the skull away from the body—a scene that paleontologists decided to put on display when the Smithsonian National Museum of Natural History fossil halls were renovated.

But this isn't to say that *T. rex* was merely a scavenger. That idea goes back to the early 1990s and, despite its popularity in television documentaries and magazine articles, was never accepted by most paleontologists. In fact, *T. rex* was both a hunter and a scavenger, much like carnivores today. The dinosaur possessed a number of predatory adaptations—such as binocular vision, allowing the dinosaur to pinpoint prey instead of scoping things out in the two-dimensional world most herbivores experienced—and, naturally, a tyrannosaur would not pass up a free meal if it happened across one. It's really the proportion of fresh prey versus carrion that *Tyrannosaurus* ate that remains unknown, but the dinosaur may have been like the spotted hyena of eastern Africa. Even though spotted hyenas are often cast as scavengers, some populations get upward of 90 percent of their diet through hunting. Their jaws—also capable of crushing bone—are just as good at bringing down prey as chewing through a carcass, meaning that they can really make the most of anything they find on the landscape. The ability to scavenge so exquisitely is not a weakness or indicative of a lack of predatory prowess. It's an adaptation that can see such carnivores through lush times when prey is plentiful as well as harder times when skin and bones might be all that there is on the menu. The fact that *Tyrannosaurus* is almost as abundant in Hell Creek as some of the large herbivores attests to this pattern, the dinosaur's chances of survival increased by its ability to turn even the largest carcasses into splinters.

T. rex undoubtedly had an outsized impact on the surrounding landscape. Paleontologists classify the carnivore as a megatheropod, which isn't just a cool-sounding title. Big animals often have such an outsized impact on the environment—what they eat, where they walk, how much dung they drop onto the soil. *T. rex* definitely falls into this category, and researchers have recently been getting a handle on why this dinosaur left such a deep mark on Hell Creek. As a species *T. rex* was around for about 2 million years. So far as we know, the dinosaur ranged from Saskatchewan to Utah, and possibly even farther south to prehistoric New Mexico. A recent study based on the fossil abundance of *T. rex* estimated that there were about twenty thousand *T. rex* around at any one time. That number is both high and low—lower than the count of herbivorous species *T. rex* ate, like *Triceratops,* but also surprisingly high for a big carnivore. Large carnivores often have large home ranges and are relatively rare, their numbers dictated by the availability of prey. But, as in other aspects of its natural history, *T. rex* wasn't quite like other carnivorous dinosaurs.

Non-avian dinosaurs changed dramatically as they grew up. Often, especially during the race to fill museums during the twentieth century, paleontologists would misidentify the juveniles of a known species as something new—that's how different the younger dinosaurs looked from their parents. *T. rex* was no exception. Young *T. rex* often had long-legged proportions that gave them a lanky look, as well as long, shallow snouts that could deliver nasty bites but weren't capable of crushing bones. Rare specimens of these young tyrants were sometimes attributed to a second Hell Creek tyrannosaur species—"*Nanotyrannus*"—but now we know that these fossils were just *T. rex* in the process of growing up. For the first fifteen years of their lives, *T. rex* filled the niche of midsized carnivore in Hell Creek and went after smaller game. This shouldered out other species of medium-sized predators, meaning that the Hell Creek ecosystem was strange compared with environments where large carnivores roam today. On the savanna of East Africa, for example, carnivores range in size from lions to bat-eared foxes, forming a continuum. In Hell Creek, there were small birds and raptors, but, with one or two exceptions, the midsized carnivore role was filled by young *T. rex*. Once those tyrants got to about fifteen years old, though, they went through a dramatic growth spurt. Teenage *T. rex* packed on the pounds, their skulls widened at the back to provide more attachment for jaw-closing muscles, and the bone-crushing dinosaurs took on the familiar, terrifying profile that turned them into rock stars with museumgoers.

I wish that I could have written a confrontation between *T. rex* and the largest prey the dinosaur ever went after, *Alamosaurus*. At the Perot Museum of Nature and Science in Dallas, for example, the skeleton of an adult *T. rex* stands next to a full-grown *Alamosaurus*—and *T. rex* looks tiny.

A. sanjuanensis was one of the largest titanosaurs, and also the last. These herbivorous dinosaurs were distant cousins of the long-necked *Apatosaurus* and *Diplodocus*, only they thrived in the Mesozoic Southern Hemisphere and got much, much bigger. Some of the largest dinosaurs ever found—like *Argentinosaurus* and *Patagotitan*—were titanosaurs. And, toward the end of the Cretaceous, some of these dinosaurs were venturing north.

In another time and place, *Alamosaurus* might have been just another enormous plant-eater, yet another ridiculously huge sauropod that vacuumed up vegetation from an incredibly lush habitat. But the dinosaur is important. During the Late Jurassic and Early Cretaceous, between 155 million and 120 million years ago, sauropods thrived in North America. But then they disappeared until about 68 million years ago. Paleontologists call this the "sauropod hiatus," millions of years when the ecology of Mesozoic North America changed. The sauropods were eventually supplanted by the ancestors of *Edmontosaurus* and *Triceratops*, the duck-billed and horned dinosaurs that could pulverize plant matter in ways sauropods could not. Yet sauropods hung on in other parts of the world, particularly the landmasses south of the equator. Eventually, the movement of the continents opened a pathway for some to return to North America, embodied by *Alamosaurus*.

To date, paleontologists have not found *Alamosaurus* fossils in the Hell Creek Formation. *Alamosaurus* is primarily known from localities in New Mexico, Texas, and Utah, and, of these, *T. rex* has been found to co-occur only in Utah. Perhaps there was something about these southern habitats that was more amenable to *Alamosaurus*, or maybe the Cretaceous ended before the dinosaur could continue spreading north. But the dinosaur came so close to the Hell Creek ecosystem that I just had to sneak it in somehow, hence the chapter coda.

Paleontologists have yet to discover an *Alamosaurus* nest. In fact, we lack direct connections between eggs, nests, and adult dinosaurs for most species. But paleontologists have found other titanosaur eggs and nests around the world, particularly in South America. From these, we know that dinosaurs like *Alamosaurus* laid round, roughly grapefruit-sized eggs. There's no evidence that adult sauropods looked after their young, and so experts hypothesize that these dinosaurs used a "lay 'em and leave 'em" strategy similar to today's sea turtles. The survival of the species was a numbers game, the environment being flooded with so many sauropod hatchlings that predators couldn't possibly eat them all. And we know a little about what some of these titanosaurs looked like from rare fossilized embryos. These delicate fossils have indicated that titanosaurs still within their eggs were already grinding their peglike teeth together, and some even had hornlike appendages on their faces that may have helped them break out of their eggs—similar to the temporary egg tooth of some modern reptiles. These dinosaurs certainly had the odds stacked against them from before they were even born, but, if they could make it to adulthood, they were some of the largest animals on the planet, stretching more than one hundred feet long from nose to tail tip.

IMPACT

In film and fiction, in museums and scientific books, dinosaurs are often depicted as being in the prime of life. When a new dinosaur is described, it's rare to see artists include injuries, signs of disease, or other difficulties dinosaurs undoubtedly suffered. That's why I introduced this chapter with an irritated *Edmontosaurus annectens*.

The great *Edmontosaurus* was one of the last hadrosaurs, commonly called duckbills. I've never really liked the title—their beaks, and certainly their teeth, don't seem very duck-like. Their look is more shovel-beaked, and a particular skull at the Los Angeles County Museum of Natural History preserves the corrugated, rough, squared-off part of the *Edmontosaurus* bill that most other fossils don't include. But my focus in this section wasn't so much on titles as on the problems an *Edmontosaurus* in Hell Creek would have to deal with.

Dinosaurs were not impervious. They were injured, got infections, broke bones, and, in point of fact, dealt with a lot of the same ailments that we do—including little parasites. Insects were evolving right alongside dinosaurs through the Mesozoic, and lice were among their numbers. Paleontologists have found some fossil lice trapped in amber, for example, but that's not all we know. By looking at the genetics of modern lice, scientists were able to discern that lice proliferated during

the Mesozoic—particularly on dinosaurs that had feathers—
and actually had to find new hosts when the K-Pg extinction
wiped out much of their food supply.

Unlike *Triceratops*, where finding multiple individuals
together is very rare, paleontologists have found bone beds
containing multiple *Edmontosaurus*. And even though the
contents of a bone bed don't necessarily indicate that the animals within interacted in life, enough *Edmontosaurus* accu-
mulations have been found that paleontologists expect that
these dinosaurs were social and moved in groups. That makes
sense given the anatomy of these dinosaurs, as well. *Edmontosaurus* was not gifted much in the way of self-defense by
evolution. The dinosaur could run fairly fast, likely faster than
T. rex, but *Edmontosaurus* did not have a spiked tail, armor
plating, horns, or other ways to actively ward off a hungry
tyrannosaur. Group living and vigilance—a tactic used by
many herbivores today—was likely much more important for
Edmontosaurus, and research into the bone structure of these
dinosaurs hints that hadrosaurs grew fast so that they might
quickly get too big for carnivores to easily take down. After
all, a thirty-foot-long *Edmontosaurus* might not have much in
the way of defensive weaponry, but it was still a big, heavy animal that could break bones while thrashing around or got a
lucky hit in with its thick, muscular tail. Much like carnivores
today, *T. rex* likely targeted the old, the sick, and the small as
preferred prey, so the most an *Edmontosaurus* could hope for
was getting too big to easily chomp, and fast.

The other dinosaur that guest-stars in this chapter is *Torosaurus latus*. Paleo diehards might be wondering why I included this controversial dinosaur in the mix. There has been

some recent debate over whether *Torosaurus* was a unique horned dinosaur that lived alongside *Triceratops* in Hell Creek or if, in fact, it represents the fully mature life stage of *Triceratops*.

Part of the problem is that *Torosaurus* is rare. While hundreds of *Triceratops* fossils have been found, only a comparative handful of *Torosaurus* specimens have turned up. That might be because *Torosaurus* is really an adult *Triceratops*, representing a time of life few *Triceratops* got to see, or because it simply was a rare animal in the same habitat, or had a different core habitat and only sometimes coexisted with *Triceratops*. Ecologists know this concept as species evenness, or the relative proportions of organisms found in a habitat. While diversity is a count of the number of species—let's say twenty dinosaurs in one ecosystem—evenness is a more important number, telling us something about who was abundant and who was rare. In this case, *Triceratops* was the most common Hell Creek dinosaur while *Torosaurus* was barely present.

What distinguishes these dinosaurs is all in the skull. While *Triceratops* has a frill of solid bone, *Torosaurus* has two large perforations, or fenestrae—meaning "windows"—in its great neck shield. More than that, *Torosaurus* fossils have twice the number of pointed epiossifications around the edge of its frill compared to *Triceratops*. Being that most known *Torosaurus* skulls are relatively large, it seemed to make sense that this dinosaur really was a big, mature *Triceratops*. But the fossil record can't be read literally. There are certain size thresholds that favor preservation for relatively large animals over the small ones. The smallest and biggest dinosaurs were rarely preserved, with dinosaurs more toward the middle of the body size distribution having a better chance. This means that young *Torosaurus*—just like very young *Triceratops*—are vanishingly rare. All the same, a *Torosaurus* recently recovered during construction outside of Denver, Colorado, is relatively small and has been nicknamed "Tiny." If this *Torosaurus* proves to be a young, still-growing animal, then that would offer strong support for the idea that *Torosaurus* is a valid species and not just a misidentified *Triceratops*.

But the real star of this chapter is the bolide itself, a chunk of rock that forever changed the planet. Part of the difficulty in telling this stone's story, though, is that very little of the bolide remains. The iridium spike in rocks found around the world—including within the impact crater itself—comes from remnants of the asteroid, to be sure, but there is no singular hunk of rock left over from impact. That's part of what led paleontologists to initially be skeptical of the impact hypothesis back in 1980—there was no smoking gun. And yet, through digging into our planet's geology and observing the pathways of asteroids in space, experts have begun to piece together an understanding of what the K-Pg bolide was like and where it came from.

Tiny fragments of the K-Pg impactor match up with a type of asteroid researchers have seen before. These are called carbonaceous chondrites. There are several different classes of these rocks, but what groups them together is that they are primarily made of unaltered dust and debris that accreted from the early—relatively speaking—days of the universe, and contain a large amount of carbon in the form of graphite, which got its name from the fact that we use it in pencils. That recognition has narrowed the field as researchers have looked for the origins of the K-Pg impactor among clues scattered through cosmic collisions that occurred millions of years before the rock even struck Earth.

One working hypothesis is that the K-Pg bolide was the remnant of a long period comet. As the name implies, these chunks of ice and rock take long, roundabout journeys through our solar system. They can take tens of millions of years to complete a single orbit around the sun. And in doing so, these comets sometimes come a little too close to the gravitational pulls of the sun and Jupiter, as the comet Shoemaker-Levy 9 did in 1992. By watching this impact, cosmologists and astronomers learned a bit more about what happens when bolides strike planets and what that powerful geologic process looks like. Then again, it's possible the asteroid might have broken apart from a much larger, older body of rock and was shunted toward Earth by the forces that dictate asteroid pathways through space. What I present here is something of a hybrid story, as the two ideas aren't mutually exclusive: a large chunk of ancient rock floated around a distant part of our solar system, and when that bolide broke apart at least one huge piece had the bad luck of heading toward Earth and striking our planet—an event so rare that the frequency is measured in terms of billions of years.

Details about the speed, direction, and angle of the asteroid are estimates based upon the latest state of the evidence. This is difficult science, a kind of geological ballistics that has to be reverse-engineered from the pattern of damage done. Perhaps future studies might modify some of the details of precisely how the rock struck Earth. All the same, we know that the asteroid was moving incredibly fast. It's a sobering point to remember, especially given how popular depictions of the asteroid impact often display the likes of *Tyrannosaurus* and *Triceratops* watching a terrible streak rip its way across the sky. This probably didn't happen. The inhabitants of the Late Cretaceous world wouldn't have seen much at all in the skies prior to impact. The event happened so fast that there wouldn't have been that telltale shooting star tail passing overhead. Likewise, the moment of impact is often shown as a terrible blinding light akin to the horrific bombing of Hiroshima and Nagasaki at the close of World War II. When the K-Pg impact is often likened to vast numbers of nuclear explosions going off at once, the metaphorical connection no doubt influences paleoart. But we don't have any indication about whether the asteroid impact created light similar to that created by a powerful bomb, much less whether such a flash would have reached the inhabitants of Hell Creek. More important to our story—and much more deadly—was the infrared pulse that followed the impact as debris began to rain down through the atmosphere. It was this glut of infrared radiation that would have been worldwide and inescapable.

The chapter's coda, on *Quetzalcoatlus,* is based upon the hypothesis that these enormous pterosaurs could have circumnavigated the globe. Standing about as tall as a modern-day giraffe and with a wingspan comparable to a small propeller plane, *Quetzalcoatlus* was one of the largest—if not *the* largest—animals ever to fly. They were truly pushing the limits of how big a vertebrate can get while still being aerodynamically capable, the fact that they leaped into the air in a pole-vault sort of fashion allowing them to circumvent some of the problems around getting enough lift to leave the ground. And once in the air, paleontologists hypothesize, *Quetzalcoatlus* and similar pterosaurs could have soared for nearly a thousand miles without having to stop. These great flying reptiles could have crossed continents, in other words, and, like birds along today's Pacific flyway, they might have split their year in different parts of the world.

But the challenging thing about studying pterosaurs is that their bones are extremely difficult to find. If you ever get the chance to see real fossil pterosaur bones in a museum, they often look like crumpled pieces of cardboard. They're so thin and full of air pockets—reducing weight to allow these animals to fly—that the bones are easily crushed during the preservation process, even when they held together long enough to be buried. In some places, all that we know of a given pterosaur species is a piece of skull or a crushed limb bone. In the case of *Quetzalcoatlus*, then, paleontologists would have to find bones of this pterosaur in different places around the planet to confirm that the pterosaur really was a world traveler. The aerodynamics of the animal indicate that the immense flyer could have done this, but finding the relevant evidence is going to be a very difficult task.

THE FIRST HOUR

For more than a century, *Ankylosaurus magniventris* has been described as a "living tank." It's difficult to come up with a better analogy for this low-slung herbivorous dinosaur. The largest of the family that bears its name—the ankylosaurs—*A. magniventris* was a rare part of the Hell Creek ecosystem that nevertheless would have frustrated many a tyrannosaur. From its snout to the end of its tail club, this dinosaur was virtually covered in bony armor.

Ankylosaurus was not the only species of its kind in the Hell Creek ecosystem. In recent years, paleontologists have recognized another from a different branch of the armored dinosaur family tree—*Denversaurus schlessmani*. The two can be told apart on the basis of their armor. Each of these large dinosaurs bore different shapes of osteoderms, or "skin bones," on their bodies; *Ankylosaurus* had a big, chunky tail club whereas *Denversaurus* did not.

Precisely what *Ankylosaurus* used that tail club for has been a paleontological puzzle for as long as the dinosaur's been known to scientists. The most obvious function would be as a defensive weapon. The club was a knob of bone about as big as the dinosaur's skull, held aloft by specialized tail vertebrae that were supported by stiff ossified tendons and a "handle" made of interlocking tail vertebrae that supported the club like the bat of a mace. The fact that some tyrannosaur species have been found with broken shinbones has led some paleontologists to hypothesize that these clubs mainly evolved as a way to beat back large and pesky carnivores. Then again, tyrannosaurs may have suffered broken bones for other reasons, just as there are other factors that could have shaped the nature of ankylosaur tails. It's possible that the clubs might have been used in combat between rivals, or even that they could have acted as decoys—fooling hungry tyrannosaurs to bite the tail clubs rather than the heads. These sorts of behavior are extremely difficult to tease out without a fossil showing the behavior in action—such as two *Ankylosaurus* with their clubs whacking each other's flanks—so the *why* of ankylosaur clubs remains an open question. The impact effects that our unfortunate *Ankylosaurus* has to cope with, though, come from different lines of evidence.

Through multiple ongoing studies of the Chicxulub crater and the surrounding area, geologists have been able to piece together a picture of the local devastation where the asteroid hit. This understanding has been hard-won—such strikes are rare, particularly when the impact melts the target rock and what had just been stone starts to act a little more like a thick soup. On top of that, the immense tsunamis created by the asteroid strike ripped up vast amounts of sediments that were then redeposited as a jumble on top of the impact site. This is not a neat, clean document of an impact as if we were looking at a crater on the moon. The devastation of the strike itself covered over the damaged and deformed rock, a testament to just how powerful this moment was.

What *Ankylosaurus* experienced in Hell Creek, however, is based upon a controversial field site in North Dakota nicknamed "Tanis." This area garnered a great deal of public attention and fanfare in the spring of 2019 when a *New Yorker* story preceded a paper documenting this particular place, which then led to accusations that the initial discoverers of the site were not given due credit and that some of the findings hyped in the press had not been published or verified. Rather than the animals preserved at the site, the initial paper focused more on the geology—and that is where the sequence of events I've pieced together came from. So far as the authors of the paper were able to discern, the impact at Chicxulub literally shook the Earth. Seismic energy from the strike radiated outward, reaching Hell Creek in several pulses within the first hour after touchdown. All this shaking, in turn, seems to have sloshed water around some of the local Hell Creek waterways. Lakes or possibly even leftover remnants of the vanishing Western Interior Seaway acted almost like swimming pools. When the earthquakes struck, the water inside those pools began to slosh and eventually overflow their banks, ripping up the bottom sediment and spilling out over the landscape in a localized flood. Small, shelly fossils testify to the violence of this event. Among the debris found at the North Dakota site are pieces of ammonites, the coil-shelled squid relatives that lived and died in the Western Interior Seaway. These fossils were from animals that lived millions of years before the time of *T. rex*, when the area was underwater. The power of the seiche waves scoured them back out of the underlying lake sediment, mixing them in with more recent fossils. If future analysis supports this sequence of events, then this site might offer a rare look into what happened shortly after the bolide buried itself in our planet.

With so much of the book focused on Hell Creek, I wanted to make sure I was paying attention to events that were occurring elsewhere. The marine reptile *Morturneria* offered an opportunity to switch the focus to Antarctica. First named in 1989, this long-necked plesiosaur was recently found to be a filter feeder. Its small, closely packed teeth were the marine reptile's equivalent of a large whale's baleen—a natural net that caught small morsels. That ecological role makes sense given that paleontologists had previously reported evidence that some long-necked plesiosaurs fed along the sea bottom. Instead of whipping their necks around like underwater serpents, which had been the classic paleoart interpretation, some Cretaceous plesiosaurs grubbed through the sand and silt on the seafloor to nab crabs, bivalves, and other invertebrate prey. Strange runnels found in ancient marine sediments, too, might have been made by plesiosaurs essentially making underwater strafing runs, moving back and forth as they plowed their jaws through the muck. Naturally, not all plesiosaurs did this. Plesiosaurs through time undoubtedly ate fish, cephalopods, and even other marine reptiles. But just as dinosaurs and mammals pioneered new niches on land during the Mesozoic, so did plesiosaurs in the seas.

The closing part of the coda, where small bits of hot debris start falling on hapless *Morturneria*, is an extrapolation. Small spherules of melted impact debris have been found all over the world, and these would have begun falling shortly after impact. Whether they reached the waters of Antarctica within the first hour post-impact is not entirely clear, but they certainly would have within the first day. And while some of this ejecta would have burned up in the atmosphere, contributing to the heat of the Paleocene's first day, the fact that spherules and shocked quartz are found in the geological record all over the planet is proof that many stayed intact long enough to make their way into the geologic record.

THE FIRST DAY

The first day of the Paleocene is a strong contender for the single worst day in the history of life on Earth. Most of this catastrophe's lasting damage occurred within the first twenty-four hours after impact, the sort of stress that is difficult to comprehend or overstate.

To be clear, there is no particular rock stratum that geologists or paleontologists can identify as belonging to the very first day of the Paleocene. Even though radiometric dating has refined the date of impact and the end of the Cretaceous, there are still error bars on any such date and any given rock layer might represent any amount of time from minutes to decades, depending on how those rock layers formed. All the same, every geological period has a boundary when one time period stops and another begins. In the case of the Cretaceous, the boundary layer containing the iridium spike, spherules, and shocked quartz is taken as the formal geologic dividing line, and the moment the asteroid touched Earth's surface is informally considered the moment when the Cretaceous ended and the Paleocene began. That means that the non-avian dinosaurs did not go extinct at the end of the Cretaceous, as is so often stated, but within the very earliest part of the Paleocene— likely within the very first day. Given that the Paleocene is part of a broader time frame called the Cenozoic, or Age of Mammals, that's why one of the most useful papers on what happened during the first day is titled "Survival in the First Hours of the Cenozoic."

The horrific events of this first day don't come from direct observations of the rock record but from extrapolating from what's seen in the rock to planetwide effects. It's clear, for example, that the asteroid strike sent a vast amount of debris into Earth's atmosphere. Drawing from that, geologists have estimated the volume of debris and how much the re-entry heat would have altered the atmosphere. The temperatures, researchers have estimated, would have been higher than an oven set to broil. The heat became so intense that dry forest debris—from leaves to dead trees—would have spontaneously ignited all over the planet. On top of that, the infrared radiation from all the burning impact debris would have created so much light that Earth's surface would have been lit up to the point where there were almost no shadows. The only way to escape this energy pulse was to shelter from it—in the water, underground, or in some other haven.

During the decades of debate and discussion about the K-Pg extinction, the tragic loss of so much biodiversity has often been treated as a protracted and long-lasting crisis. Even in museum displays and documentaries, paleoartists have often envisioned shivering and starving dinosaurs that failed to make it through the depths of the impact winter. But current evidence suggests that most non-avian dinosaurs did not even survive that long. The infrared pulse, along with the firestorms the event sparked, would have been hot enough to virtually consume all exposed organic matter on the planet's surface. There really was no place for dinosaurs such as *Tyrannosaurus* or pterosaurs such as *Quetzalcoatlus* to hide, and the same would have been true for any mammals, small reptiles, insects, plants, fungi, or other multicellular animal life out on the surface. The majority of extinction event losses happened almost immediately.

Talking about heat offered me a chance to explore dinosaur physiology and how these animals—in one fashion or another—evolved to be hot-blooded creatures. To this day, paleontologists are still unsure just how dinosaurs accomplished this. Even among living animals, a diverse group of species will have a differing physiological profile according to their needs and behaviors. Not to mention that the world is not neatly divided into endotherms or ectotherms: there are animals that can become ectothermic while they go into torpor, for instance, just as there are animals whose body temperatures remain elevated from the background but nevertheless fluctuate. What is absolutely clear, though, is that there was no animal on Earth capable of coping with the heat of the global infrared pulse. The only way to survive the hours of heat was to shelter, and much of Earth's terrestrial life was incapable of doing so.

Which brings us to a pair of creatures that, by chance, had the behaviors and adaptations needed to survive the heat pulse—the small mammal *Mesodma* and the turtle *Compsemys*. Their stories are a combination of fact and extrapolation from other sources, an estimation of what it might have been like for such survivors. Given that both *Mesodma* and *Compsemys* are found in rocks dating to the Paleocene as well as the latest Cretaceous, we can infer that there was something about their natural history that allowed them to persist through the events that killed off so many other forms of life.

In the case of *Mesodma*, we know relatively little about the entire nature of the animal. Mesozoic mammals are often known and identified by their distinctive teeth. There are two reasons for this. The first is that—unlike the great dinosaurs— mammal species had significantly different tooth shapes from one to the next. Whereas tyrannosaurs like *Gorgosaurus* and *Albertosaurus* might be impossible to tell apart or even identify from a lone tooth, fossil mammals possessed teeth with bumps, valleys, ridges, and tubercles that were specific to particular species. Find a lone molar and you can get an idea of who's around. Which brings us to the second half of the issue. Often, there's little more than a molar left. While some more complete mammal skeletons have been found in Late Cretaceous deposits—such as that of the otter-like marsupial *Didelphodon*—most of the time teeth were the only parts of the mammals to make it into the fossil record. The vagaries of preservation are often biased against smaller fossils, which break down more easily or can be more readily gobbled up by scavengers, and so much of what we know about Mesozoic mammals comes from teeth, such as the multicusped molars that have identified *Mesodma* as a multituberculate.

To fill out the nature of *Mesodma* for the purposes of this story, then, I turned to a related animal that had lived about 10 million years before impact. At a particular fossil site in Montana brimming with remains of dinosaur nests and skeletons, paleontologists found the exceptional remains of another multituberculate called *Filikomys primaveus*. These little nibblers were diggers and lived in social groups—think of ground squirrels that form prairie dog towns through the American west and you've got the right idea. These mammals lived among the dinosaurs, sometimes even gnawing the bones of the deceased terrible lizards, and so they seemed like a good candidate to help fill out the lifestyle of their later relative *Mesodma*—which was likely a burrower as well, based on the fact that it survived into the Paleocene. The idea that *Mesodma* ate baby dinosaurs when possible, however, is speculation. Paleontologists have found multituberculate gnawing damage on dinosaur bones, and a different badger-like mammal called *Repenomamus* has been found with baby dinosaurs inside its gut, but there's no direct evidence that multituberculates ate dinosaur hatchlings. Given the proximity of such an abundant food source, though, it wouldn't be surprising if the little mammals made the most of dinner at their doorstep.

That *Mesodma* would have survived in underground shelters comes from studies of forest fires. Naturalists have long noted that when natural wildfires break out, small mammals tend to either disperse and return—as shrews do—or go underground. In this case, running somewhere else wouldn't work. There truly was nowhere to hide from the awful heat. But sanctuary underground was certainly effective. Even extreme heat reaches only a few inches down into soil. Whatever organisms sheltered more than four inches below the surface were largely safe from the first day's post-impact effects.

The fate of the turtle *Compsemys* is another supposition based upon the pattern of the fossil record and what we know about turtles today. This turtle was another survivor, found both below and above the K-Pg boundary layer. In their case, however, submersion beneath lakes or ponds was much more likely. Based upon its known anatomy, *Compsemys* was very much like living snapping turtles. The sharply hooked beak of the turtle indicates that it likely ate flesh rather than plants. And while such turtles are already good at lying still on the pond bottom, slowly using up each breath as they wait for prey, biologists have also come to learn that many turtles are able to increase their bottom time by extracting oxygen through specialized cells around their cloaca. While it is impossible to know for sure whether *Compsemys* had such an ability without an exceptionally preserved specimen that includes tissues fossilized to the cellular level, the great antiquity of turtles and the fact that some Hell Creek turtles would not look out of place in a modern pond led me to project the oxygen-boost ability to ancient *Compsemys*. So long as the turtles could remain beneath the surface for a few hours, they could survive the worst of the heat. That's because water does a great job of soaking up infrared radiation in its top layers. Semi-aquatic and air-breathing organisms such as the crocodiles and turtles would have to deal with the unpleasantness of sticking their nostrils above the surface to breathe, but, so long as they stayed underwater, they'd be shielded from the deadly heat.

To drive home that these events were unfolding all over the planet, and not just western North America, I decided to focus on Late Cretaceous India. At the time, India was still an island continent slowly being pushed toward mainland Asia by continental drift. When the landmass would arrive, millions of years later, the collision would push up the Himalayas and allow creatures that had been evolving in relative isolation to mingle with those in Asia. During the time in our story, though, India was still an island where doomed dinosaurs dwelled. Among those unfortunate animals discovered so far is *Jainosaurus septentrionalis*—a long-necked sauropod dinosaur distantly related to North America's *Alamosaurus*. While not the biggest of the big, this little-known dinosaur nevertheless grew to lengths exceeding sixty feet from nose to tail. That would have been far too big to hide from the asteroid's aftereffects, and, given the global reach of the infrared pulse, *Jainosaurus* went extinct at about the same time as the non-avian dinosaurs of Hell Creek.

THE FIRST MONTH

Trying to determine what life was like a month after impact is a difficult task. That's because a month is a human construct, a totally arbitrary delineation of time that makes sense to us but wouldn't have mattered a jot to the early Paleocene creatures. Nevertheless, checking in at the one-month mark offers an opportunity to look at some of the bigger-picture changes beginning to sweep over the world.

I felt strongly that every chapter of this book needed a living protagonist, some sort of species that might help tell the story of the Cretaceous and Paleocene worlds. In this case, I chose *Acheroraptor temertyorum*—a recently named raptor that stalked small prey through the Hell Creek forests. To date, very little is known of this particular species. The initial paper is based upon parts of the upper and lower jaw—although there has been some discussion in paleontological circles as to whether the larger *Dakotaraptor* is in fact *Acheroraptor* all grown up. Paleontologists need to find more of both forms to be sure, particularly bones that can be microscopically sampled to see what the mature forms of both *Acheroraptor* and *Dakotaraptor* looked like.

Much of what paleontologists expect of *Acheroraptor*, then, comes from related animals. The little Hell Creek raptor was a relative of the famous *Velociraptor* from Mongolia, part of a group of dinosaurs called dromaeosaurids that had curved killing claws on the second toe of each foot. As indicated by tracks, these dinosaurs held the claws off the ground when not in use and left V-shaped tracks wherever they went. When

Acheroraptor spotted prey, however, it's likely that this small carnivore leaped upon its meal-to-be and sunk in its killing claws similar to the way that red-tailed hawks and other predatory birds do today. And given that the relatives of *Acheroraptor* show evidence of extensive feathery coats, it's probable that *Acheroraptor* had long arm feathers that the dinosaur could flap to help pin down prey—or, in other situations, to help provide their feet with more grip while going up inclines. Extensive feathery coats gave many dromaeosaurs aerodynamic abilities, even on the ground, that made them more effective hunters of the Cretaceous forests.

That some small *Acheroraptor* might have survived the infrared pulse is entirely speculation. To date, no one has found conclusive evidence that any non-avian dinosaurs survived for more than a few hours into the Paleocene. The few reported cases in the literature can be explained by crocodiles with theropod-like teeth, fossil bones that were inadvertently excavated by erosion and reworked into Paleocene strata, or mistakes in radiometric dating. Nevertheless, it's not outside the realm of possibility that someone might eventually find a small non-avian dinosaur that was able to cut out a living for a little longer than its larger relatives. Paleontologists have already found evidence that some herbivorous dinosaurs created burrows, for example. The behavior was in the dinosaurian repertoire. And while it's unlikely that *Acheroraptor* made such burrows, it's entirely possible that some of these small dinosaurs could have made use of abandoned burrows or underground homes created by other species. Small mammals or turtles might have dug out burrows of suitable size and depth for a plucky *Acheroraptor* or other small dinosaur to hide in, buying the animal at least a little more time. As paleontologists continue to excavate and study the Late Cretaceous and Early Paleocene strata of the world, it would not come as a total surprise if some small non-avian dinosaur species had been able to squeak by on the basis of good luck.

But if a raptor survived, it would face a desolate world that would only become more hostile. The extinction heat pulse effectively destroyed the world's forests and leveled the Hell Creek ecosystem. The small prey *Acheroraptor* or any other small predator relied on would be incredibly scarce. This might explain why toothed birds also disappeared around this time. The geochemical effects of the impact would gradually make matters worse.

An asteroid impact is not only about the destructive force of rock plowing into the planet's surface. The K-Pg impactor struck so deep, and so harshly, that it vaporized some of the target rock. Geologists have been able to discern that some of those underlying rocks were rich in sulfur compounds, and studies of human-caused climate change have taught experts what those chemicals can do in the atmosphere. Our understanding of the Paleocene impact winter comes from our knowledge of threats to our modern world. In this case, aerosolized sulfates are known to reflect the energy coming to us from the sun. This contributes to global cooling. Geologists have estimated that the sheer amount of sulfur compounds spread through the upper atmosphere and buffeted about by high-altitude winds would have triggered a long-lasting impact winter—further intensified by soot from fires and impact debris clouding the air.

At a glance, you might expect that the extreme eruptions of the Deccan Traps only contributed to this problem. Intense volcanic eruptions have been pinpointed as the cause of global weather changes before, such as the "year without a summer" when the eruption of Mount Tambora in 1815 in Indonesia caused annual temperatures in North America and western Europe to drop. In fact, the eruptions were so massive that some geologists previously hypothesized that they—rather than the asteroid impact—were the primary drivers of the end-Cretaceous extinction. But recent research has altered the story. First, the Deccan Traps did not erupt once but several times—both before and after the impact. Additionally, a 2020 study that estimated how much of the planet would be habitable for non-avian dinosaurs around the time of the K-Pg disaster found that the volcanic eruptions actually mitigated the effects of the sulfates by spewing global warming gases into the air. This not only exonerated the Deccan Traps from their role as extinction suspects but asserted that the impact winter would have been significantly worse had it not been for the well-timed eruptions in prehistoric India.

For the chapter's coda, I wanted to reinforce the point that mammals did not make it through the mass extinction unscathed—and that the story of recovery was not the same everywhere around the planet. Multituberculates are part of that story.

So far as paleontologists have been able to track, the earliest multis evolved during the major flowering of mammal diversity during the Jurassic. This was the time when mammals were no longer just insectivores, but various groups evolved to be like beavers, flying squirrels, aardvarks, and more. Multis took on a rodent-like role during that time, becoming one of the most successful mammal groups of all time. As a group, they even survived the K-Pg disaster, with species like *Mesodma* setting the stage for their future.

Yet, despite their success during the Age of Dinosaurs, multis never recovered their previous prominence. In fact, by about 10 million years after impact multis were extinct. (The date depends on which fossil mammals belong to the group, but even by the most generous estimates these mammals disappeared tens of millions of years ago.) Multis lasted for over 100 million years—the most successful group of mammals that has ever existed—and yet they declined when mammals were supposed to be on the upswing. The origin of rodents might have had something to do with that.

Precisely when animal groups began to diversify is a major point of concern for paleontologists and evolutionary biologists. It's important to know who evolved from whom and when. Those patterns are critical for understanding major events like mass extinctions. In the case of birds, for example, researchers have recently found that some bird lineages—such as waterfowl—evolved and started to make their mark on the world *before* the K-Pg impact rather than evolving in the aftermath, as previously thought. The same holds true for mammals. While the Paleocene world saw mammalian evolution take off, some modern mammal groups had their origins in the Late Cretaceous—and among these possible early risers were rodents.

So far, no one has found rodent fossils from the earliest Paleocene. The existence of such a beast is suggested by genetic data and the molecular clocks that estimate when particular mammal groups appeared, but there's not yet a fossil to stick a name to the creature in this chapter. All the same, anatomists expect that the earliest rodents were not like rats or mice but instead resembled a roly-poly living species called the mountain beaver found high in the mountain meadows of the Pacific Northwest. Their teeth and other anatomical features mark them as rodents, and they spend much of their time gathering forage from a few feet around their burrows, where they live alone. It's impossible to tell if the earliest rodents shared these habits, but the anatomy of the mountain beaver at least provides an outline of what the world's first rodents were like.

What allowed rodents to establish themselves, and eventually spread the world over, might have simply been luck. For decades paleontologists have debated why rodents were able to evolve and proliferate when multis were already filling the same niche. In some parts of the world, in fact, it seems that rodents don't really move in until the multis are on their way out. But recent analysis of who lived where and when suggests that the multis in Asia didn't bounce back strongly like their counterparts on other continents. The multis there were hit hard, and, based on the combination of available genetic and fossil information, rodents were among the evolutionary newcomers on the continent. Rodents were able to fill the niches left open by the multis, effectively shouldering them out. In addition, for reasons not yet understood, it seems that rodents and their relatives were able to recover faster than multis and other forms of more ancient mammals. Maybe there was some difference in reproduction that allowed rodents to leave more offspring, more quickly, gradually pushing multis to the sidelines and eventually into extinction. More fossil finds and analyses will undoubtedly alter the story, but for now the fall of the multis seems tied to the rise of the rodents.

ONE YEAR AFTER IMPACT

Imagine a winter that lasted for years. Based upon calculations of the various global warming and cooling compounds tossed into the air by the K-Pg impact, the Deccan Traps, and the Earth's natural processes, geologists estimate that the impact winter lasted for as long as three years before it began to dissipate.

Through the lens of Deep Time, three years is nothing. Three years would not even register as a blip when compared to the billions of years Earth has existed. But for life, those three years were critical. Had the impact winter lasted much longer, it's possible that more of the Cretaceous survivors would have expired and the world's recovery during the earliest days of the Paleocene would have been even more difficult.

Part of the story is told through plants. Based upon which plants are found in the layers prior to and after the strike, the world's plants went through their own mass extinction. And just like the animals, plants were affected by both heat and cold. The post-impact heat pulse destroyed vast swaths of the world's forests. Unable to move, the only mature plants that had a chance of survival in the first hours of the Paleocene were those that were somehow shielded from the terrible heat and were not consumed by the raging forest fires that followed. Much like the non-avian dinosaurs, most plants had no escape from the stresses of that first day.

The way plants reproduce gave some of them an advantage. Root systems dug deep into the soil survived, as did seeds and nuts. The world's underground seed bank had the potential to make the world green again. But there was a catch. The impact and the fires cast huge amounts of particulates and debris into the air. Combined with the effects of the aerosolized sulfur compounds, sunlight was dimmed to the point that ecosystems began to fail. Evidence from the oceans, for example, indicates that the only algae that were able to survive were those that could also eat other organisms—holding them over until the sunlight came back and photosynthesis could begin again. And on land, studies of *Mesocyparis* trees from China—a surviving plant that was also present in western North America at the time—indicate that these trees evolved larger seed cones, perhaps as an enticement for animals to eat them and spread those seeds over a wider area.

This connection between plants and surviving animals may have been critical. From the time birds were recognized as living dinosaurs in the late twentieth century, paleontologists have wondered what allowed birds to persist while all their non-avian relatives—every single species and lineage—went extinct. What made birds special? The answer involves digging deeper, beyond the fact that birds survived and asking which birds survived.

Flying away wouldn't have done much good during the Paleocene heat pulse. There was nowhere to fly to, and birds attempting to flap away would be exposed to the broiling heat that likely finished off the flying pterosaurs. The birds that survived the first day likely did so by burrowing or finding rock shelters along bodies of water. (While there were waterbirds by this time, they could not hold their breath like turtles or crocodiles could.) And even as the heat ebbed, the surviving birds faced a world that was nearly wiped clean. Like unfortunate *Acheroraptor*, any birds that relied on hunting or feeding on small prey probably starved. The decline in insects—documented by changes in leaf damage between the Cretaceous and Paleocene—and the mass extinctions of mammals, lizards, and snakes leave little doubt that any small, carnivorous dinosaurs that survived into the Paleocene soon ran out of food. Under stress, the world was one that favored herbivores and omnivores.

The fact that only beaked birds survived the K-Pg extinction, and are the only birds found in the fossil record for the past 66 million years, provides a significant part of the answer. Beaks had evolved among some bird lineages long before the end of the Cretaceous, part of new bird species filling up Cretaceous habitats. Studies of how fast birds grew compared with dinosaurs and investigations of how birds lost their ancestral teeth have underscored how the ancestors of modern birds switched away from their carnivorous diets to become more reliant on plants, from soft vegetation to fruits and seeds. Part of this shift involved the evolution of a muscular gizzard that could effectively "chew" food as it passed through the digestive system, as well as the origin of a crop birds could use to store excess food to get through harsh times. These fortunate anatomical quirks didn't evolve after the impact, in other words, but had been there long before, and the birds that survived the impact winter were the ones that were able to pluck up seeds and other hard-to-crack foods from the devastated forests. The hypothesis is still new and may develop additional wrinkles as paleontologists discuss the idea, but to date it is the most compelling argument for why beaked birds survived while small toothed birds and raptor-like dinosaurs did not.

I decided to focus the chapter's coda on a reptile with a Hell Creek connection, albeit in a different place. *Thoracosaurus neocesariensis* was a gharial-like crocodile that lived in watery habitats between the Late Cretaceous and the Paleocene. Somehow, likely by sheltering underwater, they were able to survive. And there is a particular fossil site in southern New Jersey where beautifully preserved *Thoracosaurus* skulls and skeletons have been found among the scattered remains of mosasaurs, sea turtles, various crocodiles, and other marine fossils. Precisely what this site represents is difficult to interpret. Many of the fossils are found in specific fossil layers, all mixed up together. It's possible that some of the bones of Cretaceous creatures were exposed and reworked into the Paleocene layers, mixing together organisms that lived at different times. But it's also possible that the vast marine boneyard represents a macabre moment from the mass extinction—a mass death assemblage of creatures that were rapidly killed in the impact's aftermath. With that in mind, and knowing that *Thoracosaurus* is one of the species that made it into the Paleocene, I decided to imagine one of the surviving crocs swimming over the sunken graveyard of creatures that it would have tried to eat or escape from just one year before. The fact that some *Thoracosaurus* specimens found in this area are immense, potentially exceeding twenty feet in length, hints that these reptiles may have spent a short time as apex predators of the early Paleocene coasts.

ONE HUNDRED YEARS AFTER IMPACT

When does a mass extinction end? That's a difficult question to answer. The incompleteness of the fossil record doesn't help here. A paleobiologist might note that a mass extinction is constrained by the time when species are becoming extinct faster than the background rate—the constant ecological shifting as some species die out and others have populations sprout off into new forms while the ancestral stock disappears. When considering the earliest part of the Paleocene, however, we don't have the resolution to track the tempo of life. In fact, we may never.

As wonderful as it can be, the fossil record can also be a frustrating thing. In the case of the K-Pg extinction, it's been incredibly difficult to find fossils of organisms that lived just before the disaster—who was really around in that last day of the Cretaceous. Often, which species were present is estimated based upon which fossils are found highest up in the geologic section. The closer a fossil is to the boundary line, the more likely it is that the species was around on the day of impact. In the case of non-avian dinosaurs, for example, finds like a *Triceratops* skull buried within three feet of the boundary layer is a pretty good indication that this dinosaur was around.

But there are two complicating factors here. The first is the fossil record itself. The fossil record is not in any way a complete record of life on Earth. It is a record of fortuitous burials—organisms or their traces that were of just the right size, present in just the right conditions, to be buried rapidly and left undisturbed, a fraction of which were brought to the surface and an even smaller fraction of which are discoverable by scientists. And this is where a paleontological phenomenon called the *Signor-Lipps effect* comes into play.

You can do this experiment at home if you have a bunch of little plastic dinosaurs or animal toys lying around, like the kind you get in a grocery store peg bag for a dollar. Take those toys in hand, close your eyes, and throw them on the floor so that they scatter. What you've just made is a facsimile of the fossil record, vestiges of life scattered through time. Now here's where the phenomenon comes into effect. Pick an orientation for your strata column—a bottom part that is older and an upper part that is younger. Now, with a piece of string or a ruler or even your imagination, draw a side-to-side line across that stack of time. That's the equivalent of a geologic boundary, the designation geologists make between one time period and another. These choices aren't quite arbitrary, but they are based upon the available rock layers and if there are any major changes in life between them. But that's not the important part. The important part is that for any given boundary in the geologic record, organisms are likely to "disappear" as they approach that boundary. This is just the nature of the fossil record. Look at your column. Let's say a little *Pteranodon* figure landed a few inches short of the boundary layer, and there's not another one any closer. If you were a geologist reading this evidence, then, you might say that *Pteranodon* went extinct well before the boundary and was not present when the extinction happened. Only, this might not be true. It can be hard to tell whether a species truly goes extinct or if it just isn't preserved for some reason. Paleontologists even have a term for the latter—a *Lazarus taxon* is an organism that was thought to be extinct but reappears at a later time after an apparent break in the fossil record. The *Metasequoia* trees of Hell Creek are one such case. The trees were named as fossils in the mid-twentieth century and were found, still alive, in the forests of China a few years later. Often, declaring a fossil species extinct requires more than just absence of fossils but some explanation of what happened: Did that species evolve into another, did the habitat change, or did something else happen?

But the second factor is not an artifact of science. It's something that was unique to Hell Creek. When ecologists of the twentieth century studied acid rain, they realized its connection to pollution and sulfates in the air. With this in mind, geologists have been able to confidently predict that acid rain would have been among the post-impact effects in the early Paleocene. Following from that, studies of small, shelly organisms called ostracods—round little crustaceans called seed shrimp for their shell-like casings—have indicated that acid rain might have effectively erased some of the slowly forming fossil record. While the acid rain wouldn't have stung on contact, it still had the potential to acidify water and soil to the point that it destroyed some of the remnants of the Hell Creek ecosystem, perhaps explaining why dinosaur bones are so hard to find in the layers around the boundary. The fires of the first day of the Paleocene obliterated much of the Cretaceous world, and then acid rain took care of some of the remainder. Add that to the incomplete nature of the fossil record and the time around impact is very difficult to approach directly.

But acid rain was not an inescapable scourge. In fact, some of Hell Creek's ponds and waterways were underlain with limestones that formed as part of the more ancient Western

Interior Seaway. These rocks created a buffer and neutralized the acidity of the rain, and this offered a reprieve for amphibians. That by itself solves something of a K-Pg mystery.

Unlike other vertebrates, amphibians don't show evidence of a mass extinction during the K-Pg. This might seem odd. There was incredible heat, an impact winter, and acid rain, the latter of which can sometimes affect the soft, sensitive amphibians that are tied to water and often respirate through their skin. The fact that frogs seem to enter the Paleocene unfazed led some experts to suggest that maybe some aspects of the impact model were incorrect or not as deadly as assumed. But the fact that the base rock of the Hell Creek ecosystem, and likely others, buffered the effects of acid rain helps provide an explanation for why amphibians were able to hop over the boundary. With that in mind, I selected *Eopelobates* as our representative amphibian—an ancient relative of the spadefoot toad that was hopping around during the Paleocene. While little about the behavior of this amphibian is known, I based my characterization on the habits of modern amphibians and tried to constrain my speculation to the parts of the story that would highlight the broader ecological changes.

The chapter's coda is based upon fossil evidence from New Zealand. As with the other vignettes, I wanted to bring a little attention to a place far away from Hell Creek and give some sense of how this event affected the entire planet. In this case, New Zealand provides a critical piece of evidence of a fern spike following the K-Pg disaster. These events are still seen today, especially in the aftermath of catastrophes like eruptions, where ferns are very quick to colonize disturbed habitats. Paleontologists have found evidence that this phenomenon was in action during prehistory, too, and the strata of New Zealand have been particularly informative. Whereas ferns made up about a quarter of the forest's plants prior to the end of the Cretaceous, in the aftermath prehistoric fern species make up about 90 percent of the earliest Paleocene plants, an initial wave that fades back once the taller forest plants began to re-establish themselves.

ONE THOUSAND YEARS AFTER IMPACT

What was transpiring in what was once the Hell Creek eco-system a thousand years after impact is something of a mystery. Radiometric dating that far into the past is not yet precise enough to give us a direct picture. Instead, I pieced together the contents of this chapter based upon what we know about the time before the extinction and after, looking to which species survived and which went extinct to come up with a speculative idea of what might have been happening.

The small snake *Coniophis* seemed like a fitting candidate for this chapter because the reptile is among the few that survived from the Cretaceous into the Paleocene. There was a mass extinction of lizards and snakes—just as there was for dinosaurs, mammals, birds, and other forms of life—but this tiny ribbon of a reptile was one of the few that persisted through the worst conditions. Given the small size of *Coniophis*, I assume that some would have survived in underground burrows where temperatures would have been cooler. Likewise, the idea that *Coniophis* could have slowed down in cooler months is based on the behavior of living snakes that go through their own form of hibernation—called brumation—during the winter. There is some debate over whether the earliest snakes evolved as swimmers or burrowers, but recent evidence tilts that scale toward the burrowing side and I favored that scenario in my tale.

The idea that mass extinctions are not primary drivers of biodiversity has come from recent, data-heavy studies of the fossil record's patterns. Even though ecosystems certainly change after mass extinctions—sometimes dramatically so—there's no indication that a mass extinction will actually generate the evolution of novelty just by clearing the field. This ties into the idea of adaptive landscapes or the hypothesis that there are a particular number of niches or forms organisms can take. Under this view, species get nudged toward adaptive "peaks" that exist in an environment, with in-between forms going extinct. It's been an influential model in evolutionary biology, but I think the argument has it backward. Niches are created by organisms rather than existing to be filled by whatever species gets there first. Following the patterns of the fossil record, for example, it's clear that—in general—the times of greatest biodiversity and disparity of form do not coincide with recovery from mass extinctions. Instead, life really took off during the Carboniferous when organisms began to live in a new setting—on land—and new interactions drove the origin of many novel species and forms. Following the K-Pg mass extinction, there's no sign that the disaster stimulated life to evolve many new species. Rather, it took time—millions of years—for interactions between survivors to create novel ecosystems that themselves created more space for evolutionary novelty.

Selecting a character for the opposite phenomenon, the inevitability of extinction, was difficult. We don't know which species, if any, perished around the thousand-year mark. Looking at the list of which species survived and which died out, however, I felt that the lizard *Palaeosaniwa* was an appropriate choice. I could see a way that juveniles of the species might have survived in underground refuges only to emerge into a world with few resources. This may not have happened. For all we know, the lizards could have died out on the first day, or within ten years, or a hundred thousand years. The fossil record is not a literal transcript of comings and goings. But I wanted to make the point that extinction is not synonymous with the death of the last member of a species. Ecologists are often concerned with functional extinction or when there are barriers to survival—such as too few individuals to sustain a breeding population—that cannot be surmounted. Compare the fates of *T. rex* with our hypothetical lizards. All *T. rex* likely died out at about the same time, on the first day of the Paleocene. They were too large to hide or escape from the infrared pulse. But smaller *Palaeosaniwa* could have survived, at a lower population density. The carnivorous lizards would have had difficulty finding food, meaning that their numbers may have declined until they reached a point that the populations could not recover or continue. The lizard went extinct, in other words, while members of the species were still alive.

The chapter's coda, on coccolithophores, comes from a recent study of these algae across the K-Pg boundary. Tiny, well-preserved fossils have indicated a major coccolith shift around this time, species capable of moving—and therefore chasing after smaller meals—dominating over those that photosynthesized. After the impact winter, however, the diversity and disparity of coccoliths increases again, indicating that the reduced sunlight had major repercussions for the oceans. More than that, the tale is a fitting example of how survival in times of ecological crisis works. Organisms cannot foresee or plan for the future. Instead, having a great degree of diversity—within a species or between species—increases the chances that some organisms will have traits that, by good fortune, allow them to survive major shake-ups. Some of those catastrophes—like the infrared pulse—are so unprecedented that many species will perish, but some have a natural history that allows them to persist. It was a lucky break that some coccoliths could hunt as well as photosynthesize, in other words, and that allowed ocean ecosystems to bounce back much more quickly than if the planktonic collapse had been total. This is the idea of contingency in evolutionary history: what has happened in the past constrains what's possible moving forward from that point.

ONE HUNDRED THOUSAND YEARS AFTER IMPACT

Devastating as the K-Pg extinction was, most of the extinction's destruction was over after a hundred thousand years. Forests were growing up again, Cretaceous survivors setting the stage for a new evolutionary moment in which flowering plants would become dominant over the conifers in many of the planet's forests. New species of mammals, birds, and other creatures were evolving, too, but this doesn't mean that the influence of the extinction was totally erased. This is still a world in the aftermath of disaster, Paleocene verdure to the contrary.

The surviving birds will play a critical part of the story. Molecular analyses of birds indicate that many modern lineages started to originate and evolve during the Late Cretaceous. Those beaked lineages survived into the Paleocene, where many more evolutionary options were open to them. There's no evidence that the flying pterosaurs made it to the Paleocene, so until the evolution of bats in the coming millions of years, birds are the only flying vertebrates on the planet at this time. By looking at Cretaceous fossils, and later fossils from the Eocene, paleontologists have been able to predict that the Paleocene was a critical time for bird evolution. Some of the avians that evolved during this time wouldn't look out of place in a modern forest, and, more importantly, this major evolutionary moment for birds likely gave some of them their characteristic smarts. A recent survey of bird evolution found that bird brains didn't shrink alongside their bodies during their evolution from dinosaur ancestors or during their Paleocene specialization. This means that birds wound up with bigger brains for their body size. This doesn't dictate intelligence— the same trend isn't seen in prehistoric mammals—but it does open up the possibility for the startling intelligence of birds like corvids and parrots to eventually evolve.

Naturally, the mammalian story is a critical part of this vignette. Determining what mammals were around one hundred thousand years after the Cretaceous, though, is difficult. My choice of *Baioconodon denverensis* is based on the discovery of this mammal in the Paleocene rocks outside Denver, Colorado. The rocks in the Denver Basin were recently dated as part of a broader study—which will have more influence on the following chapter—and so *Baioconodon* is a good candidate for a mammal that was trundling around the forest during this time. I chose to have this mammal munch on a spider because of an ancient mammalian perk, one that started to fade away during the Paleocene. During the Cretaceous, the earliest placental mammals were likely insectivores. Aside from their anatomy, genetic studies have hypothesized that the first placental mammals had five functional copies of a gene that makes chitinase—a digestive enzyme that breaks down arthropod carapaces. *Baioconodon* would have likely inherited a few functioning copies of these genes, especially if it was an omnivorous mammal. It was only as the surviving mammals began to split into new families with new specializations, geneticists hypothesize, that some mammals began to lose their chitin-digesting enzymes.

I also wanted to highlight a crocodylian in this section. These reptiles are often cast as living fossils, virtually unchanged since the heyday of the dinosaurs. Paleontologists have long known that this isn't true. The earliest crocs were terrestrial animals, and the Mesozoic saw a burst of crocodylian innovation. There were crocs that ran through the forest, swam through the sea, ambushed dinosaurs at the water's edge, and even had mammal-like teeth to eat a variety of foods. But any terrestrial crocs that were around on the last day of the Cretaceous were quickly driven to extinction. The only surviving species, as far as anyone can tell, were those that were semi-aquatic and could rely on the water for protection. Even though terrestrial, dinosaur-like crocs—some even with hooves—would evolve again during the Cenozoic, they would start from this basic anatomical plan. The presence of such a creature at one hundred thousand years after impact is also based on fossils from that time period found in the Denver Basin, encased in obscuring concretions.

The coda for the chapter deals with the disappearance of one of my favorite fossil lineages—the ammonites. These circular cephalopods were a mainstay of the oceans from before the first dinosaurs were eating beetles. They thrived throughout the Mesozoic, often playing a critical role in the seas. From fossil gut contents we know at least some ammonites ate plankton, building their bodies and producing vast amounts of young based on the foundation of ocean ecosystems. Ammonites, in turn, became prey for many marine reptiles and other Mesozoic creatures. But even though ammonites had survived mass extinction events before, by the end of the Cretaceous they seem to have been fewer in number. No one is sure why. And yet, there is some evidence that ammonites were able to survive one hundred thousand years or so after impact. Early Paleocene ammonite fossils indicate that these cephalopods somehow struggled through. But just as with the multituberculates, surviving the K-Pg disaster did not guarantee success in the Paleocene world. When the planktonic ecosystems crashed during the impact winter, a great number of ammonites were killed off. Some paleontologists speculate that ammonites often ate their own young when those offspring were tiny enough to be part of the ocean's planktonic profile. Those straggling lineages were battered by the extinction, and it's possible that they either faded away or had to compete for scant resources with ancient nautilus species, which fared better after the destruction. This only deepens the mystery of why the ammonites disappeared. They didn't all vanish immediately on the first day of the Paleocene, but somehow they could no longer keep up with life's rhythm in the aftermath.

ONE MILLION YEARS AFTER IMPACT

The key to this chapter, and in fact part of the inspiration for this book, comes from Colorado's Denver Basin. During the time I was starting to write this book a team of paleontologists reported on an absolute bonanza of fossils that documented the comings and goings of plants and animals during the first million years of the Paleocene.

Some of the fossils found at Corral Bluffs, such as *Baioconodon*, were from lower down, closer to the Cretaceous. But higher up were hardened capsules of rock containing skulls and other fossils. These fossils are the basis for *Eoconodon* and the crocodylian in the chapter. Both have been found at Corral Bluffs, and, so far, *Eoconodon* is the largest mammal recorded from the site. While Paleocene mammals might not normally be counted as rock stars even among paleontologists, to have such a timer of what life was like so soon after the K-Pg catastrophe is incredibly important to understanding life's recovery.

Eoconodon was not a specialized carnivore. While I've given my version some cravings for meat, the mammal was still an omnivore—possibly something like a big coati or raccoon. It ate more flesh than some other mammals of its time, but it was not the Paleocene equivalent of a jaguar or wolf. It certainly had a powerful bite, though. Skulls of *Eoconodon* show broad cheekbones where powerful jaw muscles threaded through the opening, providing enough power to tackle a variety of foods.

The decision for *Eoconodon* to eat bean pods is speculative, and used here to illustrate a point. One of the most significant finds at Corral Bluffs was not a mammal, but the fossil remains of an early legume. These plants—which include beans—are important because legumes are energy-rich plants. A mammal could get more nutrition from them than simply chewing leaves. The evolution of legumes, as well as new angiosperms that were decked with fruit or dropped nuts onto the forest floor, provided Paleocene mammals with energy-rich fodder, the fuel mammals would need to start becoming larger and more ecologically specialized. In fact, as a study published in 2021 indicated, even though dinosaurs constrained Mesozoic mammals from getting too big, it was competition between mammal species in this limited space that prevented a greater variety of forms from evolving. Now that mammals could evolve to be big and small, carnivorous and herbivorous, and everything in between, their story was rapidly changing.

Unlike *Eoconodon,* however, *Purgatorius unio* was not found at Corral Bluffs. Instead, this small mammal has been found in the rocks of Hell Creek itself and in the overlying Paleocene rocks of the area. The important mammal is known from bits and pieces rather than skeletons, but by following the anatomical trail, experts have cast *Purgatorius* as one of the earliest-known primates—founding members of our lineage that somehow survived the K-Pg extinction and made a home in the Paleocene forests.

The social habits of *Purgatorius* that I've included are

speculative. But based on the primate's ankle bones and other aspects of its anatomy, paleontologists think that *Purgatorius* lived something like a modern tree shrew. *Purgatorius* did not have binocular vision like we do, and it did not have an opposable thumb, but such a creature—skittering through the trees and nabbing insects—marks the beginnings of the lineage that we belong to. Using those inherited chitinase enzymes, *Purgatorius* munched insects—a diet that gradually changed as primates continued to evolve, split into new lineages, and spin off a particular branch called anthropoids that would eventually come to include our own species.

Acknowledgments

Books come together over what feels like geological time-scales. Even the pages look like stacked layers of stone. And sometimes the conditions of life when you start a book can change into something else by the time you finish.

I owe a debt of gratitude to Lark Willey and Foxfeather Zenkova for their encouragement as I came up with the idea that eventually became *The Last Days of the Dinosaurs*. They may be part of my personal Cretaceous, but I do not wish to dishonor that history.

Bee Brookshire has been a friend and a trusted source of advice for about as long as I've been writing. We've seen each other in moments of both success and frustration, but we've both kept tapping away at our keyboards. Having a friend who shares the struggle can make all the difference, and it was a joy to celebrate the small victories as we both tackled our respective manuscripts. I'm also grateful to Carrie Levitt-Bussian, Kit Morgan, Alex Porpora, and other close friends who urged me on and listened to my occasional writer's laments as I went about composing these pages.

I'd been mulling over the idea of an extinction book for

years, but it was a conversation with Berkeley paleontologist Pat Holroyd at the 2018 Society of Vertebrate Paleontology meeting that made me dig in further. No one has told this story in detail before, and now, half a century after the debate over the decimation of the dinosaurs started, there are enough threads to pull this story together. Pat's vote of confidence when I explained the idea helped me create the book you now hold.

I'm also indebted to my web editors at *Smithsonian Magazine* during the time I worked on this book, especially Brian Wolly, Jay Bennett, Beth Py-Liberman, and Joe Spring. Keeping up with the K-Pg catastrophe as part of my freelance beat for the website allowed me to stay on top of the latest research and helped set the stage for this book.

While they cannot read these words, the animals that prowl and pad around my home made this book possible. Jet, Hobbes, Terra, and Teddy provided plenty of wags and purrs, whether I wrote a thousand words in a day or none at all. Knowing them also helped inform some of the mammalian portions of this book. A cat or a dog can teach you a lot about being a mammal.

Naturally, any book like this needs a great agent and an equally great editor. Deirdre Mullane has been a tireless advocate for my work, whether it was sold or not, and Daniela Rapp at St. Martin's Press has turned a speculative idea to something much grander than I could have imagined. Especially during a year when the pandemic doldrums and depression ground me to a creative halt at times, I am thankful for their patience, insight, and faith that this book was one worth writing.

But most of all, I'm thankful to my girlfriend, Splash. One

of my previous books, *My Beloved Brontosaurus*, brought us together. On good days and bad, whether I felt hopeful or self-critical about my output, she'd smile and say, "I can't wait to read your book." I always believed her. She heard these stories first and saw the messy process by which they came to life. I strove to write something that could equal the writing that led us to meet in the first place.

Notes

Preface

vii "on a perfectly normal Cretaceous day 66 million years ago. . . ." P. Renne, A. Deino, F. Hilgen, et al. 2013. Time scales of critical events around the Cretaceous-Paleogene boundary. *Science* 339 (6120): 684–87.

vii "an explosive force 10 billion times greater . . ." Rocks at asteroid impact site record first day of dinosaur extinction, *UT News*, September 9, 2019. https://news.utexas.edu/2019/09/09/rocks-at-asteroid-impact-site-record -first-day-of-dinosaur-extinction/.

viii "For years, it was called the Cretaceous-Tertiary . . ." E. Molina, L. Alegret, I. Arenillas, et al. 2006. The global boundary stratotype section and point for the base of the Danian Stage (Paleocene, Paleogene, "Tertiary," Cenozoic) at El Kef, Tunisia: Original definition and revision: *Episodes* 29 (4): 263–73.

viii "A chunk of space debris that likely measured more than seven miles across . . ." L. Alvarez, W. Alvarez, F. Asaro, and H. Michel. 1980. Extra-terrestrial cause for the Cretaceous-Tertiary extinction. *Science* 208 (4448): 1095–108.

viii "The heat, fire, soot, and death blanketed the planet in a matter of hours. . . ." D. Robertson, M. McKenna, O. Toon, et al. 2004. Survival in the first hours of the Cenozoic. *GSA Bulletin* 116 (5–6): 760–68.

viii–ix "99 percent of all species that once lived are now extinct . . ." M. Novacek and Q. Wheeler. 1992. Extinct taxa: Accounting for 99.999 . . . % of the Earth's biota, in *Extinction and Phylogeny*, eds. Novacek and Wheeler (New York: Columbia University Press), 1–16.

ix "paleontologists have estimated that about 75 percent of known species . . ." D. Jablonski and W. Chaloner. 1994. Extinctions in the fossil record (and discussion). *Philosophical Transactions of the Royal Society of London B* 344 (1307): 11–17.

x "Just as the dinosaurs once benefitted from a mass extinction . . ." R. Irmis, S. Nesbitt, K. Padian, et al. 2007. A Late Triassic dinosauromorph assemblage from New Mexico and the rise of dinosaurs. *Science* 317 (5836): 358–61.

INTRODUCTION

1 "Magnolias and dogwoods shoulder their way . . ." P. Wilson, G. Wilson Mantilla, and C. Stromberg. 2021. Seafood salad: A diverse latest Cretaceous florule from eastern Montana. *Cretaceous Research* 121 (5981): 104734.

1 "A *Triceratops horridus* ambles along . . ." J. Scannella, D. Fowler, M. Goodwin, and J. Horner. 2014. Evolutionary trends in *Triceratops* from the Hell Creek Formation, Montana. *PNAS* 111 (28): 10245–50.

2 "At this time of the day . . ." N. Arens and S. Allen. 2014. A florule from the base of the Hell Creek Formation in the type area of eastern Montana: Implications for vegetation and climate, in *Through the End of the Cretaceous in the Type Locality of the Hell Creek Formation in Montana and Adjacent Areas*, eds. G. Wilson Mantilla, W. Clemens, J. Horner, and J. Hartman. (Washington, DC: Geological Society of America).

2 "A small flock of avian species will carry on their family's banner . . ." A. Balanoff, G. Bever, T. Rowe, and M. Norell. 2013. Evolutionary origins of the avian brain. *Nature* 501: 93–96.

3 "Fliers like *Quetzalcoatlus* . . ." M. Witton and M. Habib. 2010. On the size and flight diversity of giant pterosaurs, the use of birds as pterosaur analogues and comments on pterosaur flightlessness. *PLOS ONE* 5 (11): e13982.

3 "In the seas, the quad-paddled . . ." T. Ikejiri, Y. Lu, and B. Zhang. 2020. Two-step extinction of Late Cretaceous marine vertebrates in northern Gulf of Mexico prolonged biodiversity loss prior to the Chicxulub impact. *Scientific Reports* 10: 4169; T. Tyrrell, A. Merico, and D. McKay. 2015. Severity of

ocean acidification following the end-Cretaceous asteroid impact. *PNAS* 112 (21): 6556–61.

3 "Marsupial mammals will almost be wiped out in North America . . ." D. Grossnickle and E. Newham. 2016. Therian mammals experience an ecomorphological radiation during the Late Cretaceous and selective extinction at the K-Pg boundary. *Proceedings of the Royal Society B* 283: 20160256; N. Longrich, B. Bhullar, and J. Gauthier. 2012. Mass extinction of lizards and snakes at the Cretaceous-Paleogene boundary. *PNAS* 109 (52): 21396–21401.

3 "Creatures of the freshwater rivers and ponds . . ." D. Robertson, W. Lewis, P. Sheehan, and O. Toon. 2013. K-Pg extinction patterns in marine and freshwater environments: The impact winter model. *Journal of Geophysical Research: Biogeosciences* 118 (3): 1006–14.

4 "NASA keeps an eye on the sky . . ." A. Chamberlin, S. Chesley, P. Chodas, et al. 2001. Sentry: An automated close approach monitoring system for near-Earth objects. *Bulletin of the American Astronomical Society* 33: 1116.

4 "Experts have often spoken of the calamity as part of the Big Five . . ." D. Jablonksi. 2001. Lessons from the past: Evolutionary impacts of mass extinctions. *PNAS* 98 (10): 5393–98.

4 "The first extinction crisis . . ." P. Sheehan. 2001. The Late Ordovician mass extinction. *Annual Review of Earth and Planetary Sciences* 29: 331–64.

4 "The second event . . ." M. Caplan and R. Buston. 1999. Devonian-Carboniferous Hangenberg mass extinction event, widespread organic-rich mudrock and anoxia: Causes and consequences. *Palaeogeography, Palaeoclimatology, Palaeoecology* 148 (4): 187–207.

4 "Worse still was the third . . ." U. Brand, R. Posenato, R. Came, et al. 2012. The end-Permian mass extinction: A rapid volcanic CO_2 and CH_4-climatic catastrophe. *Chemical Geology* 322–323: 121–44.

5 "Following that, about 201 million years ago . . ." J. Davies, A. Marzoli, H. Bertrand, et al. 2017. End-Triassic mass extinction started by intrusive CAMP activity. *Nature Communications* 8: 15596.

6 "there were scores of species we'll never know . . ." A. Chiarenza, P. Mannion, D. Lunt, et al. 2019. Ecological niche modelling does not support

climatically-driven dinosaur diversity decline before the Cretaceous/Paleogene mass extinction. *Nature Communications* 10: 1091.

6 "When the famously cantankerous British anatomist . . ." D. Naish and D. Martill. 2007. Dinosaurs of Great Britain and the role of the Geological Society of London in their discovery: Basal Dinosauria and Saurischia. *Journal of the Geological Society* 164: 493–510.

7 "Experts in the early twentieth century . . ." B. Switek. 2013. *My Beloved Brontosaurus* (New York: FSG), 190–200.

8 "Almost everyone had an opinion. . . ." M. Benton. 1990. Scientific methodologies in collision: The history of the study of the extinction of the dinosaurs. *Evolutionary Biology* 24: 371–400.

9 "The invertebrate record showed a sharp uptick . . ." N. MacLeod. 1998. Impacts and marine invertebrate extinctions. *Geological Society, London, Special Publications* 140: 217–46.

10 "First proposed in 1980 . . ." L. Alvarez, W. Alvarez, F. Asaro, and H. Michel. 1980. Extraterrestrial cause for the Cretaceous-Tertiary extinction. *Science* 208 (4448): 1095–108.

10 "But the discovery of an enormous impact crater . . ." A. Hildebrand, G. Penfield, D. Kring, et al. 1991. Chicxulub crater: A possible Cretaceous/Tertiary boundary impact crater on the Yucatán Peninsula, Mexico. *Geology* 19 (9): 867–71.

10 "the initial impact that created the Chicxulub crater . . ." R. Worth, S. Sigurdsson, and C. House. 2013. Seeding life on the moons of the outer planets via lithopanspermia. *Astrobiology* 13 (12): 1155–65.

11 "About 35 million years ago . . ." R. Tagle and P. Claeys. 2005. An ordinary chondrite impactor for the Popigai crater, Siberia. *Geochimica et Cosmochimica Acta* 69 (11): 2877–89.

Before Impact

18 "the creep of malignant cancer . . ." S. Ekhtiari, K. Chiba, S. Popovic, et al. 2020. First case of osteosarcoma in a dinosaur: A multimodal diagnosis. *The Lancet* 21 (8): 1021–1022.

21 "a wily *Edmontosaurus* can outrun a *rex* . . ." W. Sellers, P. Manning, T.

Lyson, et al. 2009. Virtual palaeontology: Gait reconstruction of extinct vertebrates using high performance computing. *Palaeontologia Electronica* 12 (3): 1–26.

21 "Better to wait near cover . . ." D. Hone and and O. Rauhut. 2010. Feeding behaviour and bone utilization by theropod dinosaurs. *Lethaia* 43 (2): 232–44.

22 "the worst adversaries of the *rex* are the smallest. . . ." E. Wolff, S. Salisbury, J. Horner, and D. Varricchio. 2009. Common avian infection plagued the tyrant dinosaurs. *PLOS ONE* 4 (9): e7288.

22 "Her earliest ancestors . . ." S. Brusatte, M. Norell, T. Carr, et al. 2010. Tyrannosaur paleobiology: New research on ancient exemplar organisms. *Science* 329 (5998): 1481–85.

23 "The insulating fluff . . ." C. Kammerer, S. Nesbitt, J. Flynn, et al. 2020. A tiny ornithodiran archosaur from the Triassic of Madagascar and the role of miniaturization in dinosaur and pterosaur ancestry. *PNAS* 117 (30): 17932–36.

23 "The Earth oozed and suppurated . . ." T. Blackburn, P. Olsen, S. Bowring, et al. 2013. Zircon U-Pb geochronology links the end-Triassic extinction with the Central Atlantic Magmatic Province. *Science* 340 (6135): 941–45.

24 "In the Northern Hemisphere . . ." T. Holtz. 2021. Theropod guild structure and the tyrannosaurid niche assimilation hypothesis: Implications for predatory dinosaur macroecology and ontogeny in later Late Cretaceous Asiamerica. *Canadian Journal of Earth Sciences* 58 (9): 778–795.

25 "the tyrant lizard king has no real equal . . ." J. Horner, M. Goodwin, and N. Myhrvold. 2011. Dinosaur census reveals abundant *Tyrannosaurus* and rare ontogenetic stages in the Upper Cretaceous Hell Creek Formation (Maastrichtian), Montana, USA. *PLOS ONE* 6 (2): e16574.

25 "But around the age of eleven . . ." F. Therrien, D. Zelenitsky, J. Voris, and K. Tanaka. 2021. Mandibular force profiles and tooth morphology in growth series of *Albertosaurus sarcophagus* and *Gorgosaurus libratus* (Tyrannosauridae: Albertosaurinae) provide evidence for an ontogenetic dietary shift in tyrannosaurids. *Canadian Journal of Earth Sciences* 58 (9): 812–828.

26 "In the whole of North America . . ." C. Marshall, D. Latorre, C. Wilson, et al. 2021. Absolute abundance and preservation rate of *Tyrannosaurus rex*. *Science* 372 (6539): 284–87.

26 "She claps her jaws . . ." P. Senter. 2009. Voices of the past: A review of Paleozoic and Mesozoic animal sounds. *Historical Biology* 20 (4): 255–87.

27 "drawing as much nutrition as possible . . ." K. Chin, T. Tokaryk, G. Erickson, and L. Calk. 1998. A king-sized theropod coprolite. *Nature* 393: 680–82.

28 "Bones that were scraped . . ." G. Erickson and K. Olson. 1994. Bite marks attributable to *Tyrannosaurus rex*: Preliminary description and implications. *Journal of Vertebrate Paleontology* 16 (1): 175–78; D. Hone and M. Watabe. 2010. New information on scavenging and selective feeding behaviour of tyrannosaurids. *Acta Palaeontologica Polonica* 55 (4): 627–34.

31 "a *thwack* from an *Ankylosaurus* club . . ." V. Arbour and L. Zanno. 2019. Tail weaponry in ankylosaurs and glyptodonts: An example of a rare but strongly convergent phenotype. *The Anatomical Record* 303 (4): 988–98.

32 "The tiny dinosaur kicks . . ." G. Grellet-Tinner, C. Sim, D. Kim, et al. 2011. Description of the first lithostrotian titanosaur embryo *in ovo* with Neutron characterization and implications for lithostrotian Aptian migration and dispersion. *Gondwana Research* 20 (2–3): 621–29.

32 "the pointed, temporary egg tooth . . ." R. Garcia. 2007. An "egg-tooth"-like structure in titanosaurian sauropod embryos. *Journal of Vertebrate Paleontology* 27 (1): 247–52.

33 "*Alamosaurus sanjuanensis* . . ." R. Tykoski and A. Fiorillo. 2015. An articulated cervical series of *Alamosaurus sanjuanensis* Gilmore, 1922 (Dinosauria, Sauropoda) from Texas: New perspective on the relationships of North America's last giant sauropod. *Journal of Systematic Palaeontology* 15 (5): 339–64.

34 "Such giants are the dinosaurs . . ." T. Williamson and A. Weil. 2008. Stratigraphic distribution of sauropods in the Upper Cretaceous of the San Juan Basin, New Mexico, with comments on North America's Cretaceous "sauropod hiatus." *Journal of Vertebrate Paleontology* 28 (4): 1218–23.

IMPACT

37 "the pebble-like scales . . ." P. Bell. 2014. A review of hadrosaurid skin impressions, in *Hadrosaurs*, eds. D. Evans and D. Eberth (Bloomington: Indiana University Press).

39 "begin to creak with arthritis . . ." J. Anne, B. Hedrick, and J. Schein. 2016. First diagnosis of septic arthritis in a dinosaur. *Royal Society Open Science* 3 (8): 160222.

40 "Without the *Edmontosaurus* or *Triceratops* . . ." J. Gill, J. Williams, S. Jackson, et al. 2009. Pleistocene megafaunal collapse, novel plant communities, and enhanced fire regimes in North America. *Science* 326 (5956): 1100–1103.

40 "The itch is a feeling . . ." V. Smith, T. Ford, K. Johnson, et al. 2011. Multiple lineages of lice pass through the K-Pg boundary. *Biology Letters* 7 (5): 782–85.

42 "a hadrosaur might chaw rotten logs . . ." K. Chin. 2007. The paleobiological implications of herbivorous dinosaur coprolites from the Upper Cretaceous Two Medicine Formation of Montana: Why eat wood? *PALAIOS* 22 (5): 554–66.

42 "The skull bones of these dinosaurs . . ." A. Nabavizadeh. 2014. Hadrosaurid jaw mechanics and the functional significance of the predentary bone, in *Hadrosaurs,* eds. D. Evans and D. Eberth.

43 "The Oort cloud is . . ." A. Siraj and A. Loeb. 2021. Breakup of a long-period comet as the origin of the dinosaur extinction. *Scientific Reports* 11 (3803).

46 "The rock is moving fast. . . ." G. Collins, N. Patel, T. Davison, et al. 2020. A steeply-inclined trajectory for the Chicxulub impact. *Nature Communications* 11 (1480).

47 "The herd passes a young female *Torosaurus* . . ." L. Maiorino, A. Farke, T. Kotsakis, and P. Piras. 2013. Is *Torosaurus Triceratops*? Geometric morphometric evidence of Late Maastrichtian ceratopsid dinosaurs. *PLOS ONE* 8 (11): e81608.

52 "Mammals have enjoyed . . ." Z. Luo. 2007. Transformation and diversification in early mammal evolution. *Nature* 450: 1011–19.

55 "There will certainly be . . ." G. Grellet-Tinner, S. Wroe, M. Thompson, and Q. Ji. 2007. A note on pterosaur nesting behavior. *Historical Biology* 19 (4): 273–77.

56 "Water is far denser . . ." S. Humphries, R. Bonser, M. Witton, and D. Martill. 2007. Did pterosaurs feed by skimming? Physical modelling and anatomical evaluation of an unusual feeding method. *PLOS Biology* 5 (8): e204.

THE FIRST HOUR

57 "Being a living tank . . ." V. Arbour and J. Mallon. 2017. Unusual cranial and postcranial anatomy in the archetypal ankylosaur *Ankylosaurus magniventris*. *Facets* 2 (2): 764–94.

60 "Too fast to even see . . ." G. Collins et al. 2020. A steeply-inclined trajectory.

61 "All that force . . ." R. DePalma, J. Smit, D. Burnham, et al. 2019. A seismically induced onshore surge deposit at the KPg boundary, North Dakota. *PNAS* 116 (17): 8190–99.

62 "waves hundreds of feet high . . ." R. Paris, K. Goto, J. Goff, and H. Yanagisawa. 2020. Advances in the study of mega-tsunamis in the geological record. *Earth-Science Reviews* 210: 103381.

65 "What goes up . . ." D. Robertson, W. Lewis, P. Sheehan, and O. Toon. 2013. K-Pg extinction: Reevaluation of the heat-fire hypothesis. *Journal of Geophysical Research: Biogeosciences* 118 (1): 329–36.

65 "The strike almost instantly . . ." P. Hull, A. Bornemann, D. Penman, et al. 2020. On impact and volcanism across the Cretaceous-Paleogene boundary. *Science* 367 (6475): 266–72.

67 "*Morturneria* has spent . . ." F. O'Keefe, R. Otero, S. Soto-Acuna, et al. 2016. Cranial anatomy of *Morturneria seymourensis* from Antarctica, and the evolution of filter feeding in plesiosaurs of the Austral Late Cretaceous. *Journal of Vertebrate Paleontology* 37 (4): e1347570.

68 "Shells, crustaceans, and . . ." C. McHenry, A. Cook, and S. Wroe. 2005. Bottom-feeding plesiosaurs. *Science* 310 (5745): 75.

THE FIRST DAY

71 "There is no dawn . . ." D. Robertson, M. McKenna, O. Toon, et al. 2004. Survival in the first hours of the Cenozoic. *GSA Bulletin* 116 (5–6): 760–68.

72 "Little *Mesodma* . . ." Y. Zhang and D. Archibald. 2007. Late Cretaceous mammalian fauna from the Hell Creek Formation, southeastern Montana. *Journal of Vertebrate Paleontology* 27 (supp. 3): 171A.

75 "The early success . . ." C. Kammerer, S. Nesbitt, J. Flynn, et al. 2020. A

tiny ornithodiran archosaur from the Triassic of Madagascar and the role of miniaturization in dinosaur and pterosaur ancestry. *PNAS* 117 (30): 17932–36.

75 "Physiology differs . . ." M. Kohler, N. Marin-Moratalla, X. Jordana, and R. Aanes. 2012. Seasonal bone growth and physiology in endotherms shed light on dinosaur physiology. *Nature* 487: 358–61.

77 "And *Mesodma* is . . ." L. Weaver, D. Varricchio, E. Sargis, et al. 2020. Early mammalian social behaviour revealed by multituberculates from a dinosaur nesting site. *Nature Ecology & Evolution* 5: 32–37.

78 "It's a legacy of . . ." Y. Haridy, M. Osenberg, A. Hilger, et al. 2021. Bone metabolism and evolutionary origin of osteocytes: Novel application of FIB-SEM tomography. *Science Advances* 7 (14): eabb9113.

79 "sharing touch and body heat . . ." C. Gilbert, S. Blanc, S. Giroud, et al. 2007. Role of huddling on the energetic of growth in a newborn altricial mammal. *American Journal of Physiology* 293 (2): R867–76.

80 "*Compsemys* is a small . . ." T. Lyson and W. Joyce. 2015. Cranial anatomy and phylogenetic placement of the enigmatic turtle *Compsemys victa* Leidy, 1856. *Journal of Paleontology* 85 (4): 789–801.

81 "At the back of *Compsemys* . . ." S. FitzGibbon and C. Franklin. 2011. The importance of the cloacal bursae as the primary site of aquatic respiration in the freshwater turtle, *Elseya albagula*. *Australian Zoologist* 35 (2): 276–82.

83 "The initial Hell Creek fires . . ." D. Robertson, P. Lewis, P. Sheehan, and O. Toon. 2013. K-Pg extinction: Reevaluation of the heat-fire hypothesis. *Journal of Geophysical Research: Biogeosciences* 118 (1): 329–36.

84 "The heads arch back . . ." A. Russell and A. Bentley. 2015. Opisthotonic head displacement in the domestic chicken and its bearing on the "dead bird" posture of non-avialan dinosaurs. *Journal of Zoology* 298 (1): 20–29.

84 "The bones of *Jainosaurus* . . ." J. Wilson, M. D'Emic, K. Curry Rogers, et al. 2009. Reassessment of sauropod dinosaur *Jainosaurus* (=*"Antarctosaurus"*) *septentrionalis* from the Upper Cretaceous of India. *Contributions from the Museum of Paleontology, University of Michigan* 32 (2): 17–40.

86 "India has broken away . . ." S. Bardhan, T. Gangopadhyay, and U. Mandal. 2002. How far did India drift during the Late Cretaceous? *Placenticeras*

kaffrarium Etheridge, 1904 (Ammonoidea) used as a measuring tape. *Sedimentary Geology* 147 (1–2): 193–217.

THE FIRST MONTH

88 "a chorus of ferns. . . ." W. Clyde, J. Ramezani, K. Johnson, et al. 2016. Direct high-precision U-Pb geochronology of the end-Cretaceous extinction and calibration of Paleocene astronomical timescales. *Earth and Planetary Science Letters* 452: 272–80.

93 "vast amounts of salt crystals . . ." A. Deutsch and F. Langenhorst. 2007. On the fate of carbonates and anhydrite in impact processes: Evidence from the Chicxulub event. *GFF* 129 (2): 155–60; K. Pope, K. Baines, A. Ocampo, and B. Ivanov. 1994. Impact winter and the Cretaceous/Tertiary extinctions: Results of a Chicxulub asteroid impact model. *Earth and Planetary Science Letters* 128 (3–4): 719–25.

94 "In such overwhelming amounts . . ." S. Lyons, A. Karp, T. Bralower, et al. 2020. Organic matter from the Chicxulub crater exacerbated K-Pg impact winter. *PNAS* 117 (41): 25327–34.

94 "Even as the fires ebbed . . ." J. Vellekoop, A. Sluijs, J. Smit, et al. 2014. Rapid short-term cooling following the Chicxulub impact at the Cretaceous-Paleogene boundary. *PNAS* 111 (21): 7537–41.

95 "Far from Hell Creek . . ." P. Hull, A. Bornemann, D. Penman, et al. 2020. On impact and volcanism across the Cretaceous-Paleogene boundary. *Science* 367 (6475): 266–72.

96 "No single factor . . ." A. Chiarenza, A. Farnsworth, P. Mannion, et al. 2020. Asteroid impact, not volcanism, caused the end-Cretaceous dinosaur extinction. *PNAS* 117 (20): 17084–93.

101 "What lice remain . . ." R. de Moya, J. Allen, A. Sweet, et al. 2019. Extensive host-switching of avian feather lice following the Cretaceous-Paleogene mass extinction event. *Communications Biology* 2: 445.

101 "The worms . . ." K. Chin, D. Pearson, and A. Ekdale. 2013. Fossil worm burrows reveal very early terrestrial animal activity and shed light on trophic resources after the end-Cretaceous mass extinction. *PLOS ONE* 8 (8): e0070920.

102 "The forests are quieter..." N. Adams, E. Rayfield, P. Cox, et al. 2019. Functional tests of the competitive exclusion hypothesis for multituberculate extinction. *Royal Society Open Science* 6 (3).

ONE YEAR AFTER IMPACT

108 "The impact winter..." C. Tabor, C. Bardeen, B. Otto-Bliesner, et al. 2020. Causes and climatic consequences of the impact winter at the Cretaceous-Paleogene boundary. *Geophysical Research Letters* 47 (3): e60121.

109 "Of more than 130..." P. Wilf and K. Johnson. 2004. Land plant extinction at the end of the Cretaceous: A quantitative analysis of the North Dakota megafloral record. *Paleobiology* 30(3): 347–68.

109 "The plants with..." Y-M. Cui, W. Wang, D. Ferguson, et al. 2019. Fossil evidence reveals how plants responded to cooling during the Cretaceous-Paleogene transition. *BMC Plant Biology* 19: 402.

112 "Most of the earliest..." N. Brocklehurst, P. Upchurch, P. Mannion, and J. O'Connor. 2012. The completeness of the fossil record of Mesozoic birds: Implications for early avian evolution. *PLOS ONE* 7(6): e39056.

113–114 "teeth helped set the timing..." T. Yang and P. Sander. 2018. The origin of the bird's beak: New insights from dinosaur incubation periods. *Biology Letters* 14: 20180090.

115 "The ancestors of beaked birds..." A. Louchart and L. Viriot. 2011. From snout to beak: The loss of teeth in birds. *Trends in Ecology & Evolution* 26 (12): 663–73.

117 "A few flicks..." D. Larson, C. Brown, and D. Evans. 2016. Dental disparity and ecological stability in bird-like dinosaurs prior to the end-Cretaceous mass extinction. *Current Biology* 26 (10): 1325–33.

117–118 "The rise of flowering plants..." G. Sun, D. Dilcher, S. Zheng, and Z. Zhou. 1998. In search of the first flower: A Jurassic angiosperm, *Archaefructus*, from northeast China. *Science* 282 (5394): 1692–95.

119 "A beak that is good for..." A. Chira, C. Cooney, J. Bright, et al. 2020. The signature of competition in ecomorphological traits across the avian radiation. *Proceedings of the Royal Society B*. Proc. R. Soc. B 287: 20201585.

120 "Most insects seem . . ." C. Labandeira, K. Johnson, and P. Lang. 2002. Preliminary assessment of insect herbivory across the Cretaceous-Tertiary boundary: Major extinction and minimum rebound, in *The Hell Creek Formation and the Cretaceous-Tertiary Boundary in the Northern Great Plains*, eds. J. Hartman, K. Johnson, and D. Nichols (Boulder, CO: Geological Society of America Special Paper 361), 297–327.

122 "about 70 percent . . ." C. Labandeira, K. Johnson, and P. Wilf. 2002. Impact of terminal Cretaceous event on plant-insect associations. *PNAS* 99 (4): 2061–66.

123 "The crocodile swims . . ." W. Gallagher, K. Miller, R. Sherrell, et al. 2012. On the last mosasaurs: Late Maastrichtian mosasaurs and the Cretaceous-Paleogene boundary in New Jersey. *Bulletin de la Société Géologique de France* 183 (2): 145–50.

ONE HUNDRED YEARS AFTER IMPACT

128 "The rain doesn't melt . . ." J. Bailey, A. Cohen, and D. Kring. 2005. Lacustrine fossil preservation in acidic environments: Implications of experimental and field studies for the Cretaceous-Paleogene boundary acid rain trauma. *PALAIOS* 20 (4): 376–89.

129 "On the edge of a deep pond . . ." R. Estes and B. Sanchiz. 2010. New discoglossid and palaeobatrachid frogs from the Late Cretaceous of Wyoming and Montana, and a review of other frogs from the Lance and Hell Creek formations. *Journal of Vertebrate Paleontology* 2 (1): 9–20.

130 "Strange as it may seem . . ." J. Cochran, N. Landman, K. Turekian, et al. 2003. Paleoceanography of the Late Cretaceous (Maastrichtian) Western Interior Seaway of North America: Evidence from Sr and O isotopes. *Palaeogeography, Palaeoclimatology, Palaeoecology* 191 (1): 45–64.

135 "Tiny cracks start to run . . ." A. Behrensmeyer. 1978. Taphonomic and ecologic information from bone weathering. *Paleobiology* 4 (2): 150–62.

136 "A beam of sunlight . . ." V. Vajda and S. McLoughlin. 2007. Extinction and recovery patterns of the vegetation across the Cretaceous-Palaeogene boundary: A tool for unravelling the causes of the end-Permian mass-extinction. *Review of Palaeobotany and Palynology* 144 (1–2): 99–112.

ONE THOUSAND YEARS AFTER IMPACT

140 "tiny *Coniophis...*" N. Longrich, B. Bhullar, and J. Gauthier. 2012. Mass extinction of lizards and snakes at the Cretaceous-Paleogene boundary. *PNAS* 109 (52): 21396–401.

141 "over 100 million years earlier. . . ." M. Caldwell, R. Nydam, A. Palci, and S. Apesteguia. 2015. The oldest known snakes from the Middle Jurassic-Lower Cretaceous provide insights on snake evolution. *Nature Communications* 6: 5996.

141 "And they burrowed. . . ." H. Yi and M. Norell. 2015. The burrowing origin of modern snakes. *Science Advances* 1 (10): e1500743.

144 "Throughout all the ages . . ." J. Cuthill, N. Guttenberg, and G. Budd. 2021. Impacts of speciation and extinction measured by an evolutionary decay clock. *Nature* 588: 636–41.

145 "This is *Palaeosaniwa...*" S. Silber, J. Geisler, and M. Bolortsetseg. 2010. Unexpected resilience of species with temperature-dependent sex determination at the Cretaceous-Paleogene boundary. *Biology Letters* 7 (2): 295–98.

146 "Extinction doesn't occur . . ." M. Galetti, R. Guevara, M. Cortes, et al. 2013. Functional extinction of birds drives rapid evolutionary changes in seed size. *Science* 340 (6136): 1086–90.

148 "The tiny sphere . . ." S. Gibbs, P. Bown, B. Ward, et al. 2020. Algal plankton turn to hunting to survive and recover from end-Cretaceous impact darkness. *Science Advances* 6 (44): eabc9123.

ONE HUNDRED THOUSAND YEARS AFTER IMPACT

151 "This is *Baioconodon...*" C. Gazin. 1941. Paleocene mammals from the Denver Basin, Colorado. *Journal of the Washington Academy of Sciences* 31 (7): 289–95.

151 "the height of the fern spike. . . ." K. Berry. 2019. Fern spore viability considered in relation to the duration of the Cretaceous-Paleogene (K-Pg) impact winter: A contribution to the discussion. *Acta Palaeobotanica* 59 (1): 19–25.

154 "Dinosaurs kept the forest . . ." M. Carvalho, C. Jaramillo, F. de la Parra, et al. 2021. Extinction at the end-Cretaceous and the origin of modern neotropical rainforests. *Science* 372 (6537): 63–68.

157 "But the asteroid strike . . ." J. Smaers, R. Rothman, D. Hudson, et al. 2021. The evolution of mammalian brain size. *Science Advances* 7 (18): eabe2101.

158 "The ubiquitous clawhold . . ." N. Brocklehurst, E. Panciroli, G. Benevento, and R. Benson. 2021. Mammaliaform extinctions as a driver of the morphological radiation of Cenozoic mammals. *Current Biology* 31 (13): 2955–63.e4.

159 "a small bird . . ." D. Ksepka, T. Stidham, and T. Williamson. 2017. Early Paleocene landbird supports rapid phylogenetic and morphological diversification of crown birds after the K-Pg mass extinction. *PNAS* 114 (30): 8047–52.

160 "the Paleocene birds are starting . . ." D. Ksepka, A. Balanoff, N. Smith, et al. 2020. Tempo and pattern of avian brain size evolution. *Current Biology* 30 (11): 2026–36.e3.

163 "The only surviving crocodiles . . ." R. Felice, D. Pol, and A. Goswami. 2021. Complex macroevolutionary dynamics underlie the evolution of the crocodyliform skull. *Proceedings of the Royal Society B* 28 (1954).

167 "It's a tiny *Pachydiscus* . . ." M. Machalski and C. Heinberg. 2005. Evidence for ammonite survival into the Danian (Paleogene) from the Cerithium Limestone at Stevns Klint, Denmark. *Bulletin of the Geological Society of Denmark* 52: 97–111.

ONE MILLION YEARS AFTER IMPACT

171 "a distraction to the *Eoconodon* . . ." T. Lyson, I. Miller, A. Bercovici, et al. 2019. Exceptional continental record of biotic recovery after the Cretaceous-Paleogene mass extinction. *Science* 366 (6468): 977–83.

177 "The babies are kept . . ." C. Janis and M. Carrano. 1991. Scaling of reproductive turnover in archosaurs and mammals: Why are large terrestrial mammals so rare? *Annales Zoologici Fennici* 28 (3–4): 201–16.

178 "One such larva . . ." M. Donovan, P. Wilf, C. Labandeira, et al. 2014. Novel insect leaf-mining after the end-Cretaceous extinction and the demise of Cretaceous leaf miners, Great Plains, USA. *PLOS ONE* 9 (7): e103542.

179 "perched on a thin branch . . ." G. Wilson Mantilla, S. Chester, W. Clemens, et al. 2021. Earliest Palaeocene purgatoriids and the initial radiation of stem primates. *Royal Society Open Science* 8 (2): 210050.

185 "The story began long before . . ." M. Chen, C. Stromberg, and G. Wilson. 2019. Assembly of modern mammal community structure driven by Late Cretaceous dental evolution, rise of flowering plants, and dinosaur demise. *PNAS* 116 (20): 9931–40.

185 "The last common ancestor of placental mammals . . ." C. Emerling, F. Delsuc, and M. Nachman. 2018. Chitinase genes (*CHIA*s) provide genomic footprints of post-Cretaceous dietary radiation in placental mammals. *Science Advances* 4 (5): eaar6478.